EXPRESS RESIDENTIAL FIRE SPRINKLER DESIGN GUIDE

U.S. Fire Administration
Emmitsburg, MD

A Cooperative Effort of

Prince George's County, MD
Fire Department
Bureau of Engineering Services

and

NAHB Research Center, Inc.
Upper Marlboro, MD

January 1995

PROLOG

The *EXPRESS Residential Design Guide* was prepared under a Cooperative Agreement with the Federal Emergency Management Administration (FEMA) and sponsored by the U.S. Fire Administration (USFA). As a cooperative effort between the NAHB Research Center, Inc., and the Bureau of Engineering Services, Prince George's County (Maryland) Fire Department, the *Design Guide* was developed for the installer as a method to simplify and expedite residential sprinkler design and layout.

The Design Guide incorporates extensive experience concerning the installation of residential fire sprinklers in accordance with NFPA 13-D. The consensus of experience maintains that most homes can be pre-engineered following design methods contained in the NFPA 13-D standard. The *Design Guide* is intended only for use in residential applications to one and two family dwellings and townhouses and is intended to assist the user in preparing designs of residential sprinkler systems which are in compliance with the NFPA 13-D standard. In the case of any discrepancy found between the minimum design standards of NFPA 13-D and the *Design Guide,* the requirements of NFPA 13-D shall be considered as taking precedence.

The writers seek qualified and responsible critique. Please address comments to:

Program Manager
Joint Residential Fire Projects
NAHB Research Center, Inc.
400 Prince George's Boulevard
Upper Marlboro, MD 20772-8731

and

David M. Banwarth, P.E.
Chief Fire Protection Engineer
Bureau of Engineering Services
Prince George's County Fire Department
County Administration Building
Room 3100
Upper Marlboro, MD 20772

THE EXPRESS RESIDENTIAL FIRE SPRINKLER *DESIGN GUIDE*

The *Guide* provides a preliminary discussion of sprinkler coverage area, water flow, and water pressure. After this discussion, the *Guide* is divided into two parts:

Part 1: Hydraulic Worksheet

Calculations performed on this sheet enable the installer to account for pressure losses in the system, to ensure that adequate **water flow** and **water pressure** are available at the most remote sprinkler head. This information, in turn, is used to select pipe sizes for the system that will permit delivery of the required flow and pressure.

Part 2: Sprinkler Target Zones

The *Guide* eliminates the need to determine precise locations for sprinkler heads, and substitutes "target zones" within which sprinkler heads can be placed to provide adequate coverage. Part 2 provides instructions for using a templet to determine target zones for each room in the house.

By following the methods described in the *Guide,* the installer can design a residential fire sprinkler system and determine the placement of the sprinkler heads.

QUESTIONS AND ANSWERS ABOUT THIS *GUIDE*

What is the Purpose of the Guide?

The purpose of the *Guide* is to provide a simplified method for designing and laying out home fire sprinkler systems. The simplification has been achieved by pre-engineering key features of the design. Use of the *Guide* can result in cost savings in sprinkler installation.

Who Can Use The Guide?

The *Guide* is intended for use by installers who are familiar with the installation of plastic and copper piping, and who possess a basic understanding of residential sprinkler system design and installation methods.

How Does the Guide Relate to the Design Method of NFPA Standard 13-D?

The method described in the *Guide* enables installers to design systems that conform to the requirements of National Fire Protection Association (NFPA) Standard 13-D, "Sprinkler Systems in One- and Two-Family Dwellings and Manufactured Homes" (1994). Two important differences between the *Guide's* method and the current 13-D method are that the *Guide* estimates the fitting loss and does not require the installer to specify the exact layout of the piping in the system.

Loops and Grids

The *Design Guide* does not support the design of looped or gridded piping networks to supply sprinkler heads. Loops and grids may be utilized to improve the hydraulic performance of a system design, such benefits are not accounted for in this simplified design method. Installers should check with the local AHJ regarding the relationship of any fire sprinkler design method to local requirements.

How Do Commercial and Residential Systems Differ?

The conventional process for fire sprinkler system design and installation was developed in commercial, industrial, and institutional settings. Such settings generally require large, complex systems, which must be fully engineered and which require substantial design review by the local AHJ. The designs are typically based on a unique set of plans and hydraulic calculations that often require special certification.

By comparison with commercial and industrial settings, homes are small. Home sprinkler systems are much less complex. The *Guide* is designed to make the advantages of this reduced complexity available to installers.

The Guide eliminates the need for detailed special plans because the system design is not tied to precise sprinkler head locations. Instead, sprinkler heads can be placed anywhere within **target zones** that are indicated on the house plans. The *Guide* substantially reduces and can eliminate the necessity for pre-installation design review. The hydraulics and head locations can be reviewed in the field after rough installation is complete. When requirements of actual construction make it necessary to change the layout of the piping or the location of sprinkler heads, conventional designs can require the submission of as-built plans. With the use of target zones, the need for as-built plans can be eliminated.

What Materials and Tools Are Needed?

All that is needed are scaled drawings of each floor of the house, and possibly a calculator.

What Information Is Needed?

Before beginning, the following information is required:

Water Supply

- The available water pressure in the home.

- The length, size, and type of underground piping material used to supply water from the street main to the house.

Change in Elevation

- The change in elevation between the water main and the highest sprinkler in the house.

Sprinkler Specifications

- Water flow, pressure, and coverage specifications for the make and model of sprinkler head that will be used. These operating characteristics appear on the product information ("cut") sheets furnished by the manufacturer.

- Specifications for any backflow prevention devices or special valves that may be used in the system. This information also accompanies the devices.

What Kinds of Homes Are Covered by the Guide?

The *Guide* can be used to design "tree systems" for most types of single-family homes. There are two exceptions:

- Placement of sprinkler heads on sloped ceilings. Manufacturers' installation guides should be followed for this type of ceiling.

- Exceptionally large homes, or homes with unique layouts and design features. For such homes, installers should prepare a design plan in accordance with NFPA Standard 13-D.

COVERAGE AREA, WATER FLOW, AND WATER PRESSURE

Sprinkler heads require different flow rates and water pressures to cover different room areas. A typical manufacturer's specification is shown in Table 1.

Table 1
Sample Sprinkler Specification

	SINGLE-HEAD FLOW		MULTIPLE-HEAD FLOW (each head)	
COVERAGE AREA	FLOW	PRESSURE	FLOW	PRESSURE
12' X 12'	10	6.6	9	5.3
14' X 14'	10	6.6	9	5.3
16' X 16'	14	12.9	11	8.0
18' X 18'	14	12.9	12	9.5
20' X 20'	16	16.8	16	16.8
Note: This table is for sample purposes only. Refer to the specific listing criteria for the particular model of sprinkler head being utilized.				

Sprinkler heads should be chosen that require as little flow as possible for the greatest coverage area. Flow rates as low as 9 to 12 gallons per minute (gpm) for a coverage area of 14' x 14' are available and constitute a desirable range.

Coverage Area

In the method described in the *Guide,* the term Coverage Area designates a **single sprinkler rating** representing the greatest coverage that any individual sprinkler head on the system will have to achieve. This Coverage Area dictates the **flow rate and the pressure for the system.** Selection of this rating is therefore the first step.

Manufacturers' coverage-area specifications for sprinkler heads typically run from 12' x 12' or less, to 20' x 20', as shown in Table 1.

Room Width

The *Guide* uses the **width of the room** as the principal dimension for determining the Coverage Area. The width is defined as the measurement of the shorter side of the room.

Here is the basic rule:

> *The greatest room width that does not exceed the greatest coverage area rating of the sprinkler heads, will determine the Coverage Area for the system being designed.*

In the following example, the house contains four rectangular rooms:

Room #1 : width, 12 feet, length 14 feet
Room #2: width, 15 feet, length 19 feet
Room #3: width, 18 feet, length 18 feet
Room #4: width, 26 feet, length 30 feet

Let us assume that the maximum Coverage Area for the sprinkler heads being used in the system is 20' x 20'. The width of Room #3, 18 feet, comes closest to the maximum rating of the heads without exceeding it. We therefore select a Coverage Area of 18' x 18' for the system.

Now let's take the rooms one by one.

- **Room #1.** Since the Coverage Area that we have chosen is greater than either dimension of this room, only one sprinkler head is required.

- **Room #2** has a width of 15 feet and a length of 19 feet. Since the length exceeds the maximum reach of our 18-foot-by-18-foot Coverage Area, a second head will be needed to provide full coverage in this room.

- **Room #3's** 18-foot length and width both fit our Coverage Area. One sprinkler head will be sufficient.

- **Room #4's** width and length both exceed the maximum reach of the sprinkler head. A **second row of sprinklers** will be required for this room. Each row will contain two sprinklers.

Room Width and Coverage Area

Table 2 summarizes the relationship between Room Width and Coverage Area. Where room widths exceed the maximum for the sprinkler head being used, the Coverage Area should be selected in accordance with the table, with the understanding that two rows of sprinkler heads are required.

Table 2
Room Width and Coverage Area

ROOM WIDTH (any length room)	COVERAGE AREA
to 12' **or** 21' - 24'	12' x 12'
to 14' **or** 25' - 28'	14' x 14'
to 16' **or** 29' - 32'	16' x 16'
to 18' **or** 33' - 36'	18' x 18'
to 20' **or** 37' - 40'	20' x 20'

6

For example, consider a house with an 17-foot x 24-foot living room and a 32-foot x 40-foot basement.

For the 17-foot-wide living room, the Coverage Area selected is 18 x 18 feet. The room requires a single row of two sprinklers to cover its 24-foot length. Figure 1 shows the target zones within which the sprinklers can be placed.

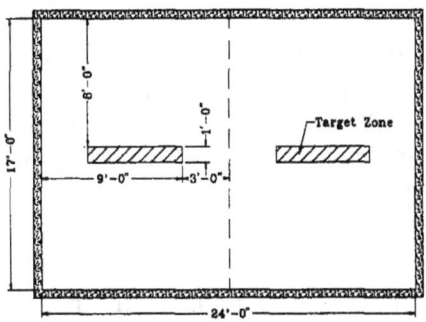

Figure 1
Living Room with Target Zones

For the basement, we find 32 feet under **Room Width** in Table 2. The table shows that the Coverage Area will be 16 feet, which falls within the 18-foot Coverage Area that we have already selected. However, two rows of sprinklers will be required to achieve 32-foot coverage. Each row will contain three sprinklers, to reach the full length of 40 feet. Figure 2 shows the target zones within which the sprinklers can be placed.

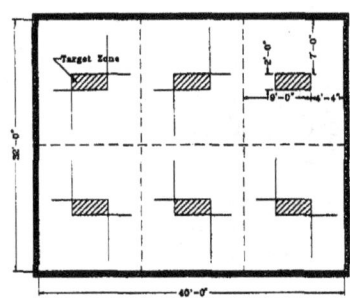

Figure 2
Basement with Target Zones

Design Water Flow (DWF) and Pressure

Two system operating characteristics, the **Design Water Flow (DWF)** and the **Water Pressure,** are based on the selected Coverage Area of the sprinkler.

Single Sprinkler Head

In the event that each room in the house has only one sprinkler head, then both the **DWF** and the required **Water Pressure** can be taken directly from the manufacturer's specifications for **Single-Head Flow** for the chosen Coverage Area for use on the hydraulic calculation worksheet.

As an example, suppose that all rooms have one sprinkler head each, and the Coverage Area is 18 'x 18'. Suppose, too, that the manufacturer's specifications exactly duplicate those in Table 1. The table gives the DWF of an 18' x 18' Coverage Area as 14 gallons per minute (gpm) and the pressure as 12.9 pounds per square inch (psi).

Multiple Sprinkler Heads

If there are more than one sprinkler head in any room of the house, then the **Multiple-Head Flow** and the **Single-Head Pressure** from the manufacturer specification, and are used on the Hydraulic Calculation Work Sheet.

- The **Multiple-Head Flow Rate** is used because it is possible that two heads in a given room with two or more sprinklers may be activated simultaneously.

- The **Single-Head Pressure** is used because this makes it possible to perform one calculation to determine adequate pressure and flow for all heads on the system, regardless of the number of heads in any individual room.

As an example, suppose that we have selected a Coverage Area of 18' x 18', and that one room in the house has two sprinkler heads. If the manufacturer's specifications are the same as those in Table 1, then the DWF is 24 gpm (12 x 2), and the Pressure is 12.9 psi.

Maximum Flow Rate

The maximum flow rate for which the *Guide* can be used is 32 gpm. Greater flows will benefit from more detailed design procedures than those described in the *Guide*.

PART 1: THE HYDRAULIC WORKSHEET

The Hydraulic Worksheet is used to determine the available pressure for the piping in the sprinkler system. This available pressure provides the basis for selecting types and sizes of piping that will permit the required pressure and water flow to be delivered to the farthest sprinkler head.

There is a drop in pressure between the tap of the public main and the farthest sprinkler head. This drop stems from four sources:

·   devices in the system that impede the water flow;

·   pressure loss as the water flows through the piping from the public main to the beginning of the sprinkler system;

·   rise in elevation between the tap of the public main and the highest sprinkler head in the house; and

·   pressure loss as water travels from the system entry point to the farthest head.

The Hydraulic Work Sheet is used to determine the available water pressure and flow rate at the farthest head after accounting for all losses. Appendix B is then used to choose types and sizes of pipe that will deliver the required pressure and flow rate.

The steps for calculating pressure losses on the Hydraulic Work Sheet are as follows:

1. Room Width and Coverage Area

Enter in **1A** the largest Room Width that does not exceed the greatest coverage area rating of the sprinkler heads. Enter the selected **Coverage Area in 1B.**

2. Sprinkler Head Specifications

For the Coverage Area in lB, fill in the following information from the manufacturer's specification sheet that is provided with the sprinkler heads:

  2A: Single-Head Flow Rate;

  2B: Single-Head Pressure; and

  2C: Dual-Head Flow Rate.

3. Design Water Flow (DWF) and Design Pressure

If all rooms in the house will have only one sprinkler head, enter the Single-Head Flow Rate (2A) in 3A.

If there will be more than one sprinkler head in any room of the house, enter the Dual-Head Blow Rate (2C) in 3B, and multiply by 2.

The applicable entry—3A or 3B x 2— is the system's **Design Water Flow (DWF).** Enter it on **Line 1.**

The Single-Head Pressure from 2B is used for the **Design Sprinkler Pressure.** Enter it on **Line 2.**

4. Water Pressure at the Public Main

This is the pressure in the public main of the local water supply system. The figure can be obtained from the utility. Enter it on Line 3.

5. Pressure Losses Caused by Devices

These losses are totalled as follows:

A. Backflow Prevention Device; Check Valve

If a backflow prevention device is present, secure the pressure loss at the DWF, from the manufacturer's specifications. Enter it on **Line 4.**

If the system employs a check valve rather than a backflow prevention device, Table 3 can be used to determine the pressure loss. Enter it on **Line 4.**

Table 3
Pressure Loss in Devices

DWF up to →	PRESSURE LOSS (psi)											
	18 gpm			22 gpm			26 gpm			32 gpm		
Size (in.)	meter	gate valve	check valve	meter	gate valve	check valve	meter	gate valve	check valve	meter	gate valve	check valve
5/8	9	-	-	14	-	-	18	-	-	-	-	-
3/4	4	1	2	8	1	3	9	2	5	-	-	-
1	2	0	1	3	0	2	3	0	2	4	1	3
1 1/4	-	0	0	-	0	1	_	0	1	-	0	2
1 1/2	0	0	0	1	0	1	2	0	1	2	0	1

If the system has neither type of device and employs a direct connection to provide flow in the system enter "0" on **Line 4.**

B. Water Meter

Determine the size of the system's water meter, and enter it in **5B.**

10

Using Table 3, find the pressure loss for this size meter at the system's Design Water Blow (Line 1). Enter this figure on **Line 5.**

C. Valves

Enter the number of gate or ball valves in 5C. Multiply by the figure that appears under the system's Design Water Flow in Table 3. Enter this figure on **Line 6.**

6. Pressure Losses in Underground Supply Piping

In 6A and 6B, enter the sizes and types of underground water service piping that extends from the public main to the house.

Appendix A provides Pressure Loss Tables for various Design Water Flows, for different types of pipe of various sizes and lengths.

- Choose the table which matches the underground pipe.

- Go down the pipe lengths in the left-hand column and choose the first one that is equal to or greater than the length of pipe that is being checked.

- Run across the column to the system's Design Water Flow to find the pressure loss.

- Enter this number on **Line 7A.**

Repeat the process for an additional pipe section, and enter the number on **Line 7B.**

7. Elevation Pressure Loss

An additional factor affecting pressure is the difference in elevation between the tap point at the water main and the highest sprinkler head in the system. Enter the difference in elevation at 7, and divide by 2. Enter the result on **Line 8.**

8. Sum of Losses and Pressure at the Farthest Head

Add Lines 2, 4, 5, 6, 7A, 7B, and 8. Enter on **Line 9.**

9. Available Pressure for Piping

To determine the available pressure for piping, subtract Line 9 from Line 3 and enter the result on **Line 10.**

10. Selection of Pipe Type and Size

Complete section 10 by using the tables in Appendix B to make pipe selection(s).

Choose the appropriate table for the DWF that you have entered on **Line 1.**

Choose the pressure loss in the vertical left-hand column that is equal to or less than the **Available Pressure for Piping** on **Line 10.**

Running across the table, choose a single pipe type or a combination of pipe types that will extend to the farthest sprinkler head.

For example, Figure 3 shows a portion of the 18 gpm Table in Appendix B. If the Available Pressure for Piping on Line 10 is 22 psi, the table shows the available piping options. These options indicate the maximum pipe run that can be accommodated by each type of pipe.

All lines except the first assume the use of two different types of piping in the maximum-length run. Select one length from Section A and one from Section B. Their combined length must accommodate the length of the run to the farthest sprinkler head.

The highlighted row in Figure 3 shows the maximum allowable length for a combined run of one-inch copper Type M and one-inch CPVC pipe. The copper portion of the run can be as long as. 50 feet, and the CPVC part of the run can be as long as 101 feet.

ALLOWABLE INSIDE PIPE LENGTHS AT 18 GPM   DESIGN WATER FLOW (DWF)

CHOOSE 1 ROW USING - ONE COLUMN FROM PIPE SECTION A PLUS ONE COLUMN FROM PIPE SECTION

18 GPM	INSIDE PIPE SECTION A		INSIDE PIPE SECTION B						
	CPVC OR CU (M) 1 1/4"	1"	CU (M) 1"	CU (M) 3/4"	CPVC 1"	CPVC 3/4"	CPVC 3/4 S"	PB 1"	PB 3/4"
15	-	-	97	27	123	41	14	50	15
15	25	-	87	24	111	37	13	45	13
15	50	-	78	22	99	33	12	40	12
15	75	-	69	19	88	29	10	35	10
15	-	25	72	20	91	30	11	37	11
15	-	50	47	13	60	20	7	24	7
15	-	75	22	6	28	9	3	11	3
20	-	-	129	36	164	55	19	67	19
20	25	-	120	33	153	51	18	62	18
20	50	-	110	31	141	47	16	57	17
20	75	-	101	28	129	43	15	52	15
20	-	25	104	29	133	44	15	54	16
20	-	50	79	22	101	33	12	41	12
20	-	75	54	15	69	23	8	28	8
25	-	-	161	45	206	68	24	83	24
25	25	-	152	42	194	64	23	78	23
25	50	-	142	40	182	60	21	74	21
25	75	-	133	37	170	56	20	69	20
25	-	25	136	38	174	58	20	70	20
25	-	50	111	31	142	47	16	57	17
25	-	75	86	24	110	36	13	44	13

(Left vertical label: AVAILABLE PRESS)

Figure 3
Allowable Inside Pipe Lengths at
18 GPM Design Water Flow (DWF),

HYDRAULIC WORKSHEET

1. ROOM WIDTH AND COVERAGE AREA

    A. Room Width: _____ ft.

    B. Coverage Area: _____ ft. x _____ ft.

2. SPRINKLER HEAD SPECIFICATIONS

    A. Single-Head Flow Rate: _____ **gpm.**

    B. Single-Head pressure: _____ psi.

    C. Dual-Head Flow Rate: _____ gpm.

3. DESIGN WATER FLOW (DWF) AND DESIGN PRESSURE

    A. If all rooms have only one sprinkler head:

        DWF (from 2A): _____ gpm.

    B. If more than one head in *any* room:

        DWF (From 2C): _____ **x 2 =** _____ gpm.

Design Water Flow (A or B above):	Line 1: _____ **gpm**
Design Sprinkler pressure (From 2B):	Line 2: _____ psi

4. WATER PRESSURE AT THE PUBLIC MAIN   Line 3: _____ psi

5. PRESSURE LOSSES CAUSED BY DEVICES

    A. Backflow prevention Device; Check Valve   Line 4: _____ psi

    B. Water Meter Loss

        Water Meter Size: _____

        pressure Loss
        (Use DWF on Line 1, and Table 3)   Line 5: _____ psi

    C. Gate or Ball Valve Loss
        (Use DWF and Table 3)
        _____ X _____ psi =   Line 6: _____ psi
        No. Valves Loss

6. PRESSURE LOSSES IN UNDERGROUND SUPPLY PIPING

Find the Pressure Losses based on the DWF on Line 1
and Tables in Appendix A.

A. Underground Section #1 Piping

$$\frac{\underline{\hspace{1cm}}, \underline{\hspace{1cm}}, \underline{\hspace{1cm}}}{\text{Size} \quad \text{Type} \quad \text{Length}} \text{ ft: Pressure Loss} =$$

Line 7A: _____ psi

B. Underground Section #2 Piping

$$\frac{\underline{\hspace{1cm}}, \underline{\hspace{1cm}}, \underline{\hspace{1cm}}}{\text{Size} \quad \text{Type} \quad \text{Length}} \text{ ft: Pressure Loss} =$$

Line 7B: _____ psi

7. ELEVATION PRESSURE LOSS

Difference in elevation between water main tap point and
highest sprinkler (if the sprinkler head is lower, the
number is negative):
_____ **/ 2 =**

Line 8: _____ psi

8. SUM OF LOSSES AND SPRINKLER PRESSURE

$$\frac{\underline{\hspace{0.8cm}}}{\text{Line 2}} + \frac{\underline{\hspace{0.8cm}}}{\text{Line 4}} + \frac{\underline{\hspace{0.8cm}}}{\text{Line 5}} + \frac{\underline{\hspace{0.8cm}}}{\text{Line 6}} +$$

$$\frac{\underline{\hspace{0.8cm}}}{\text{Line 7A}} + \frac{\underline{\hspace{0.8cm}}}{\text{Line 7B}} + \frac{\underline{\hspace{0.8cm}}}{\text{Line 8}} =$$

Line 9: _____ psi

9. AVAILABLE PRESSURE FOR PIPING

$$\frac{\underline{\hspace{1cm}}}{\text{Line 3}} - \frac{\underline{\hspace{1cm}}}{\text{Line 9}} =$$

Line 10: _____ psi

10. SELECTION OF PIPE TYPE AND SIZE

Use the appropriate Table in Appendix B, based on the
DWF, Line 1. Find the Available Pressure for Piping,
Line 9, in the Table's left-hand column. Select the
piping type(s) and size(s).

INSIDE SECTION A: $\underbrace{\underline{\hspace{1cm}}, \underline{\hspace{1cm}}, \underline{\hspace{1cm}}}_{\text{Type} \quad \text{Size}}$ ft. maximum straight length

INSIDE SECTION B: $\underbrace{\underline{\hspace{1cm}}, \underline{\hspace{1cm}}, \underline{\hspace{1cm}}}_{\text{Type} \quad \text{Size}}$ ft. maximum straight length

14

PART 2: SPRINKLER TARGET ZONES

In the traditional sprinkler head placement method, the exact locations of the sprinkler heads are indicated on a set of sprinkler plans and must be fixed according to the hydraulic design. A typical placement is shown in Figure 4.

However, considerations that arise during construction often prevent installation at design location shown on the drawings. Deviations from the locations indicated on the plans can cause coordination problems with trades, and can result in re-inspection and the submittal of "as-built" plans.

Instead of exact locations, the *Guide* substitutes **target zones** within which the sprinkler heads can be placed. Placement of the sprinkler heads anywhere within the target zones meets the requirement of NFPA 13-D to comply with the manufacturers' coverage area specifications. This method simplifies planning and design. A typical target zone created by use of the Guide is shown in Figure 5.

The Target Zone Templet

In Appendix C, two templets are provided, one for plans drawn to 1/4" = 1' scale, and one for plans drawn to 1/8" = 1' scale. Figure 6 shows a reduced copy of the 1/4" templet.

The templets have four features:

- **Along the left-hand edge:** A scale rule marked in feet, for taking measurements on the drawing.

- **On the lower portion:** Squares representing sprinkler coverage areas ranging from 12' x 12' to 20' x 20'.

- **In the middle:** An eight-foot scale, used to establish eight-foot minimum spacing between sprinklers as required by NFPA 13-D.

- **In the upper right-hand corner: The scales** that are used to lay out the target zones on the plans.

Limitations of Template Use

There are two limitations on use of the templet:

- Do not use the templet to determine target zone locations on other than flat ceilings.

- In areas that are subject to heat from such sources as stove tops, fireplaces, furnaces, and hot water heaters, the target zone may have to be modified in accordance with the manufacturers' installation guides for procedures relating to such areas.

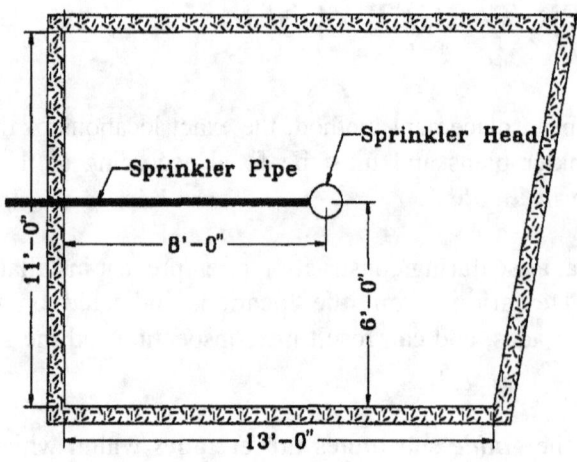

Figure 4
Traditional Placement of Sprinkler Heads

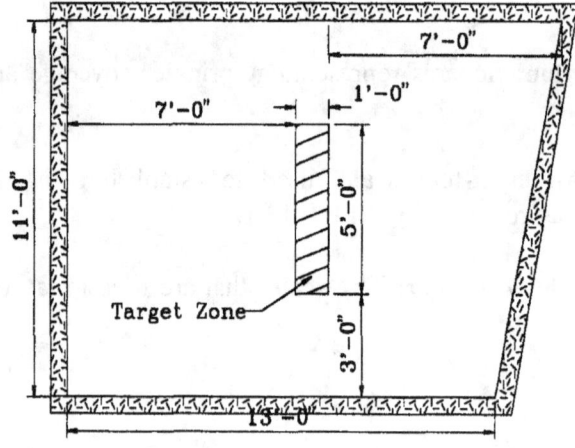

Figure 5
Typical Target Zone Based on 16' x 16' Coverage

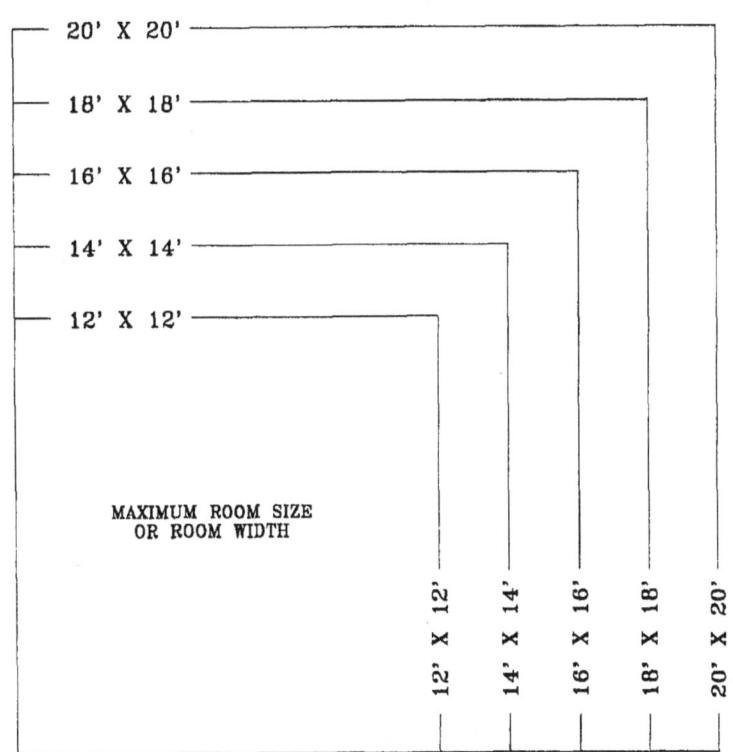

RESIDENTIAL SPRINKLER
TARGET ZONE TEMPLET
1/4" = 1' scale

EDGE LINE

EDGE LINE

12' X 12'
14' X 14'
16' X 16'
18' X 18'
20' X 20'

MINIMUM DISTANCE
BETWEEN SPRINKLERS

8'-0"

20' X 20'

18' X 18'

16' X 16'

14' X 14'

12' X 12'

MAXIMUM ROOM SIZE
OR ROOM WIDTH

12' X 12'

14' X 14'

16' X 16'

18' X 18'

20' X 20'

Figure 6
Template for the Creation of Target Zones (not to seale)

17

Two Initial Steps

Two steps must be taken before measurements begin:

- The applicable templet-- 1/4" or 1/8" --must be copied onto a transparency of the type used for overhead projections. With this transparency, house plans can be seen beneath the templet.

- The two sides of the target zone scale that face the edges of the sheet, must be trimmed along the lines marked "edge line." This is shown in Figure 7.

Coverage Area

Use of the templet to lay out target zones is based on the **Coverage Area** that has been selected for the system. This figure appears on the Hydraulic Work Sheet, Item 1B.

Creating a Target Zone

Creating a target zone by means of the templet is shown in Figures 8 and 9. In this example, the Coverage Area selected for the system is 16' x 16', and the room in which the sprinkler head is to be installed is 13' x 15'.

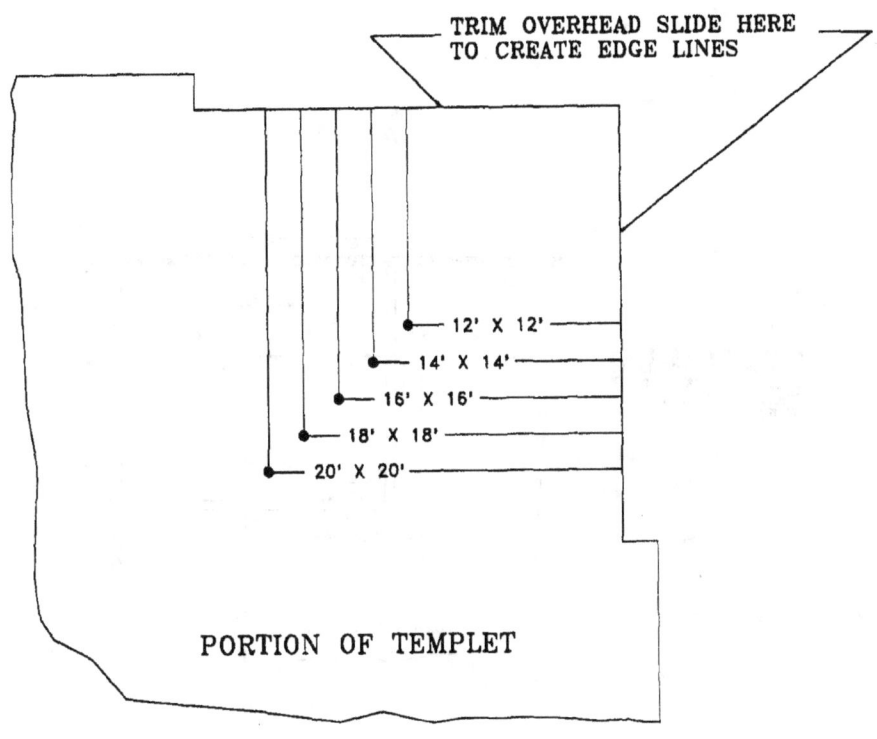

TRIM OVERHEAD SLIDE HERE
TO CREATE EDGE LINES

12' X 12'
14' X 14'
16' X 16'
18' X 18'
20' X 20'

PORTION OF TEMPLET

Figure 7
Trim Target Zone at Edge Line

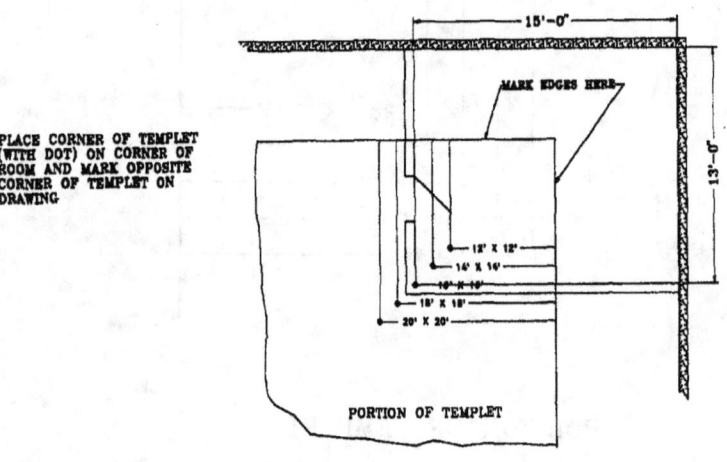

PLACE CORNER OF TEMPLET
(WITH DOT) ON CORNER OF
ROOM AND MARK OPPOSITE
CORNER OF TEMPLET ON
DRAWING

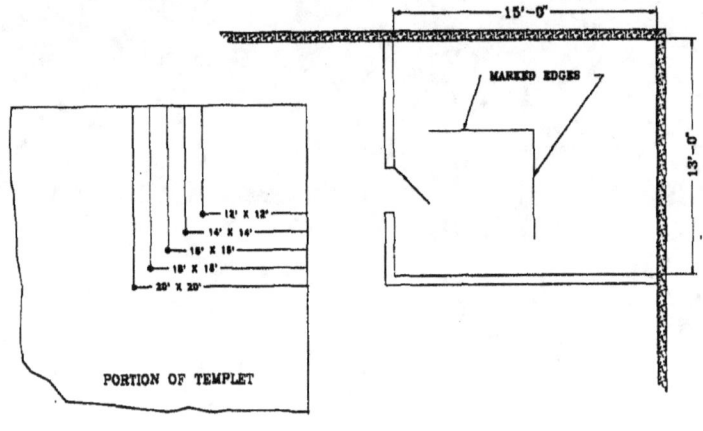

Figure 8
Creating One Corner of the Target Zone

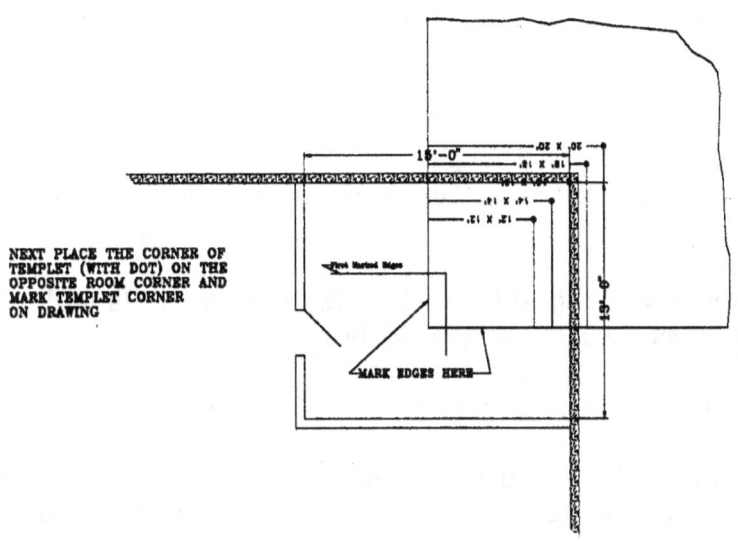

NEXT PLACE THE CORNER OF
TEMPLET (WITH DOT) ON THE
OPPOSITE ROOM CORNER AND
MARK TEMPLET CORNER
ON DRAWING

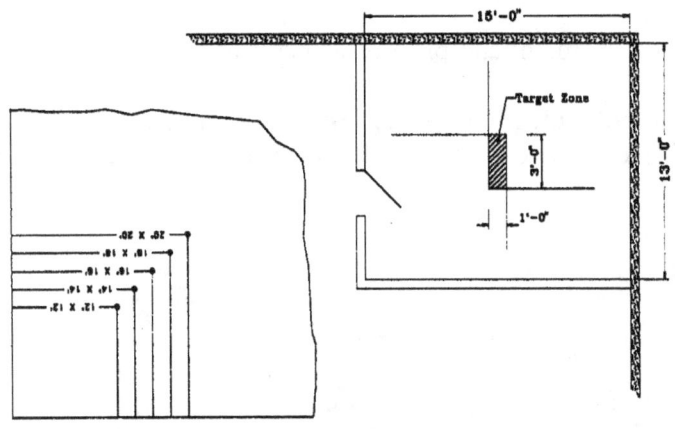

Figure 9
Creating the Second Corner of the Target Zone

The steps are as follows:

- To match the dimensions of the Coverage Area, the 16' x 16' scale in the upper right corner of the templet is employed.

- Place this scale on one corner of the room on the plans, with the dot in the corner, as shown in Figure 8. Trace the comer of the scale edges onto the drawing.

- Place the scale on the diagonally opposite sides of the room with the dot in the comer, as shown in Figure 9. Trace the comer of the scale edges onto the drawing.

- The two right-angle markings will overlap in an area in the middle of the room. **This is the Target Zone.**

Three Types of Rooms

Rooms will fall into three general types. Procedures, illustrated in Figure 10, for each type are as follows:

Type 1 - Rooms whose length and width are both less than the width of the Coverage Area. This type of room will require only one sprinkler head.

Use of the template to locate the target zone is described above.

Type 2 - Rooms whose width is less than the width of the Coverage Area, but whose length exceeds the length of the Coverage Area. This type of room will require two sprinkler heads.

The steps are as follows:

1. Divide the room in half along its length on the plans.

2. Treat each half as a separate room, and proceed as described above.

Type 3 - Rooms whose length and width both exceed the length and width of the Coverage Area. This type of room will require two rows of sprinkler heads, with each row having two heads.

The steps are as follows:

1. Divide the room in half along both its length and width.

2. Treat each of the four sections as a separate room, and proceed as above.

Using 16' x 16' Spacing:

Type 1:
Room Width and
Length Less Than
Coverage Area

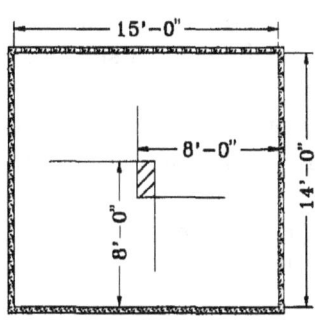

Type 2:
Room Width Less Than
Coverage Area,
Room Length Greater,
Than Coverage Area

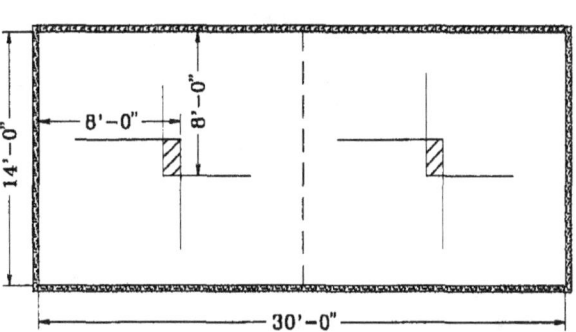

Type 3:
Room Width and
Length Greater
Than Coverage
Area

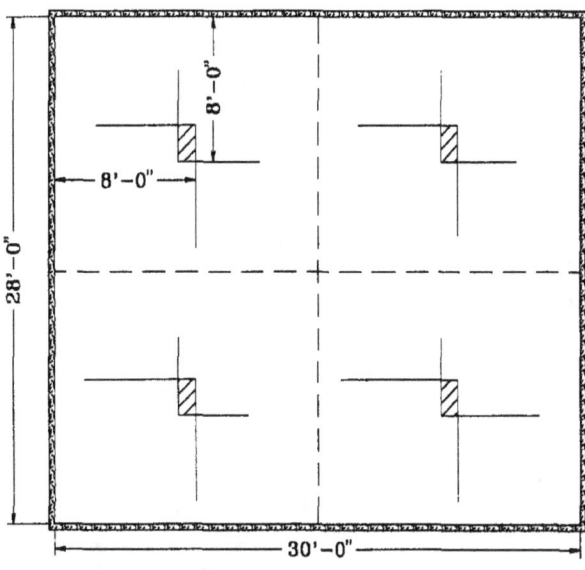

Figure 10
Target Zones in Different Room Types

Sidewall Sprinklers

To determine target zones for sidewall sprinklers, proceed as follows:

Select the wall on which the sprinkler head will be placed.

Place the templet on one side of a scale plan of the wall and trace one edge of the template onto the drawing as shown in Figure 11.

Move the templet to the opposite edge, and trace the edge onto the drawing. This will create a pair of parallel lines on the drawing. The space between them is the target zone.

For walls with more than one sprinkler, divide the wall in half and treat each half as an independent wall section.

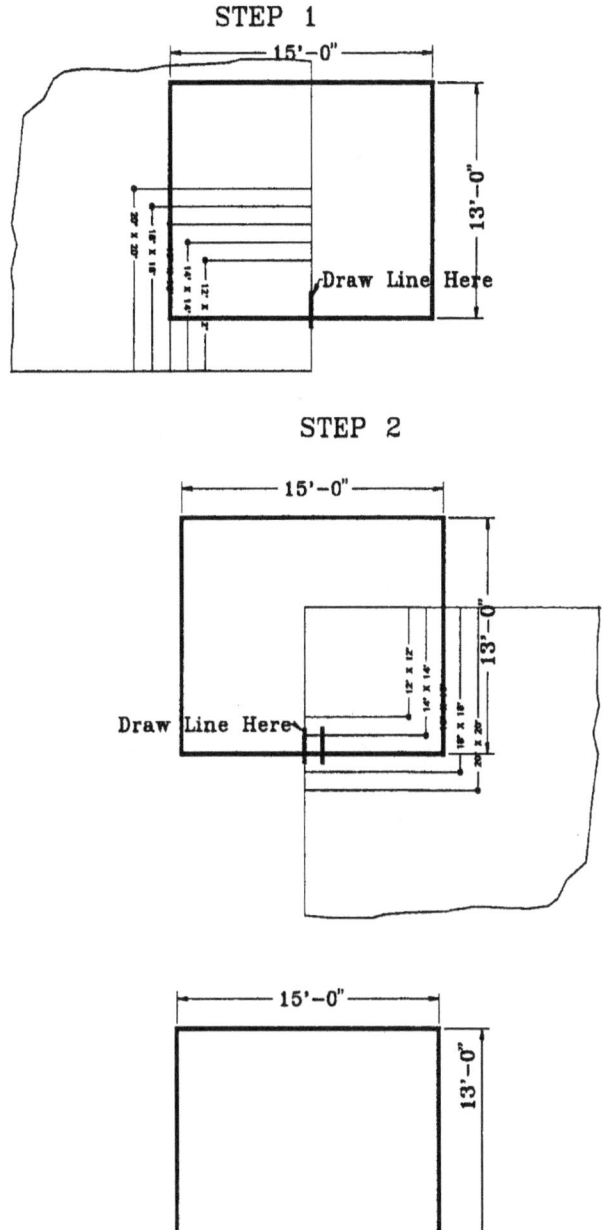

Figure 11
Target Zone for Sidewall Sprinklers

APPENDIX A
PRESSURE LOSS TABLES
UNDERGROUND PIPING

PRESSURE LOSS TABLE (psi)

3/4" COPPER PIPE (any type)

DESIGN WATER FLOW (gpm)

LENGTH OF PIPE (ft)	9	10	11	12	13	14	15	16	17	18	19	20	21	22	23	24	25	26	27	28	29	30	31	32
10	1	1	2	2	2	2	3	3	3	4	4	5	5	5	6	6	7	7	8	8	9	10	10	11
20	2	3	3	4	4	5	5	6	7	8	8	9	10	11	12	13	14	15	16	17	18	19	21	22
30	3	4	5	6	6	7	8	9	10	11	12	14	15	16	18	19	21	22	24	25	27	29	31	33
40	4	5	6	7	8	9	11	12	14	15	17	18	20	22	24	26	28	30	32	34	36	39	-	-
50	5	6	8	9	10	12	13	15	17	19	21	23	25	27	30	32	34	37	-	-	-	-	-	-
60	6	8	9	11	12	14	16	18	20	23	25	27	30	33	35	38	-	-	-	-	-	-	-	-
70	7	9	11	12	14	16	19	21	24	26	29	32	35	38	-	-	-	-	-	-	-	-	-	-
80	8	10	12	14	16	19	21	24	27	30	33	36	-	-	-	-	-	-	-	-	-	-	-	-
90	9	11	14	16	18	21	24	27	30	34	37	-	-	-	-	-	-	-	-	-	-	-	-	-
100	10	13	15	18	21	24	27	30	34	38	-	-	-	-	-	-	-	-	-	-	-	-	-	-
110	11	14	17	19	23	26	29	33	37	-	-	-	-	-	-	-	-	-	-	-	-	-	-	-
120	12	15	18	21	25	28	32	36	-	-	-	-	-	-	-	-	-	-	-	-	-	-	-	-
130	14	16	20	23	27	31	35	-	-	-	-	-	-	-	-	-	-	-	-	-	-	-	-	-
140	15	18	21	25	29	33	37	-	-	-	-	-	-	-	-	-	-	-	-	-	-	-	-	-
150	16	19	23	27	31	35	-	-	-	-	-	-	-	-	-	-	-	-	-	-	-	-	-	-
160	17	20	24	28	33	38	-	-	-	-	-	-	-	-	-	-	-	-	-	-	-	-	-	-
170	18	21	26	30	35	-	-	-	-	-	-	-	-	-	-	-	-	-	-	-	-	-	-	-
180	19	23	27	32	37	-	-	-	-	-	-	-	-	-	-	-	-	-	-	-	-	-	-	-
190	20	24	29	34	-	-	-	-	-	-	-	-	-	-	-	-	-	-	-	-	-	-	-	-
200	21	25	30	35	-	-	-	-	-	-	-	-	-	-	-	-	-	-	-	-	-	-	-	-

C = 150   i.d. = 0.745"

PRESSURE LOSS TABLE (psi)

1" COPPER PIPE (any type)

LENGTH (feet)	\ gpm	9	10	11	12	13	14	15	16	17	18	19	20	21	22	23	24	25	26	27	28	29	30	31	32
10		0	0	0	0	1	1	1	1	1	1	1	1	1	1	1	2	2	2	2	2	2	2	3	3
20		1	1	1	1	1	1	2	2	2	2	2	2	2	3	3	3	3	4	4	4	4	5	5	5
30		1	1	1	2	2	2	2	2	3	3	3	3	4	4	4	5	5	5	6	6	7	7	8	8
40		1	1	2	2	2	3	3	3	4	4	4	5	5	5	6	6	7	7	8	8	9	9	10	11
50		1	2	2	2	3	3	3	4	4	5	5	6	6	7	7	8	8	9	10	10	11	12	13	13
60		2	2	2	3	3	3	4	4	5	6	6	7	7	8	9	9	10	11	12	12	13	14	15	16
70		2	2	3	3	4	4	5	5	6	6	7	8	9	9	10	11	12	13	14	15	16	17	18	19
80		2	3	3	3	4	5	5	6	7	7	8	9	10	11	12	12	13	14	16	17	18	19	20	21
90		2	3	3	4	5	5	6	7	7	8	9	10	11	12	13	14	15	16	17	19	20	21	23	24
100		3	3	4	4	5	6	7	7	8	9	10	11	12	13	14	16	17	18	19	21	22	24	25	27
110		3	3	4	5	6	6	7	8	9	10	11	12	13	15	16	17	19	20	21	23	24	26	28	29
120		3	4	4	5	6	7	8	9	10	11	12	13	15	16	17	19	20	22	23	25	27	28	30	32
130		3	4	5	6	7	7	8	10	11	12	13	14	16	17	19	20	22	24	25	27	29	31	33	35
140		4	4	5	6	7	8	9	10	12	13	14	16	17	19	20	22	24	25	27	29	31	33	35	37
150		4	5	6	7	8	9	10	11	12	14	15	17	18	20	22	23	25	27	29	31	33	35	38	-
160		4	5	6	7	8	9	10	12	13	15	16	18	20	21	23	25	27	29	31	33	35	38	-	-
170		4	5	6	7	9	10	11	13	14	16	17	19	21	23	25	27	29	31	33	35	38	-	-	-
180		5	6	7	8	9	10	12	13	15	17	18	20	22	24	26	28	30	33	35	37	-	-	-	-
190		5	6	7	8	10	11	12	14	16	17	19	21	23	25	27	30	32	34	37	-	-	-	-	-
200		5	6	7	9	10	12	13	15	16	18	20	22	24	27	29	31	34	36	39	-	-	-	-	-

DESIGN WATER FLOW (gpm)

C = 150   i.d. = 0.995"

PRESSURE LOSS TABLE (psi)

1 1/4" COPPER PIPE (any type)

DESIGN WATER FLOW (gpm)

LENGTH OF PIPE ft

Length (ft)	9	10	11	12	13	14	15	16	17	18	19	20	21	22	23	24	25	26	27	28	29	30	31	32
10	0	0	0	0	0	0	0	0	0	0	0	0	0	0	0	1	1	1	1	1	1	1	1	1
20	0	0	0	0	0	0	0	0	1	1	1	1	1	1	1	1	1	1	1	1	1	2	2	2
30	0	0	0	0	1	1	1	1	1	1	1	1	1	1	1	2	2	2	2	2	2	2	3	3
40	0	0	0	1	1	1	1	1	1	1	1	2	2	2	2	2	2	2	3	3	3	3	3	4
50	0	1	1	1	1	1	1	1	1	2	2	2	2	2	2	3	3	3	3	4	4	4	4	4
60	1	1	1	1	1	1	1	1	2	2	2	2	2	3	3	3	3	4	4	4	4	5	5	5
70	1	1	1	1	1	1	2	2	2	2	2	3	3	3	3	4	4	4	5	5	5	6	6	6
80	1	1	1	1	1	2	2	2	2	2	3	3	3	4	4	4	5	5	5	6	6	6	7	7
90	1	1	1	1	2	2	2	2	3	3	3	3	4	4	4	5	5	6	6	6	7	7	8	8
100	1	1	1	1	2	2	2	2	3	3	3	4	4	4	5	5	6	6	7	7	7	8	8	9
110	1	1	1	2	2	2	2	3	3	3	4	4	5	5	5	6	6	7	7	8	8	9	9	10
120	1	1	1	2	2	2	3	3	3	4	4	5	5	5	6	6	7	7	8	8	9	10	10	11
130	1	1	2	2	2	3	3	3	4	4	4	5	5	6	6	7	7	8	9	9	10	10	11	12
140	1	1	2	2	2	3	3	3	4	4	5	5	6	6	7	7	8	9	9	10	10	11	12	12
150	1	2	2	2	3	3	3	4	4	5	5	6	6	7	7	8	9	9	10	11	11	12	13	13
160	1	2	2	2	3	3	4	4	4	5	5	6	7	7	8	8	9	10	11	11	12	13	14	14
170	1	2	2	2	3	3	4	4	5	5	6	6	7	8	8	9	10	10	11	12	13	14	14	15
180	2	2	2	3	3	3	4	4	5	6	6	7	7	8	9	10	10	11	12	13	13	14	15	16
190	2	2	2	3	3	4	4	5	5	6	7	7	8	9	9	10	11	12	12	13	14	15	16	17
200	2	2	2	3	3	4	4	5	6	6	7	8	8	9	10	11	11	12	13	14	15	16	17	18

C = 150   i.d. = 1.245"

PRESSURE LOSS TABLE (psi)

1 1/2" COPPER PIPE (any type)

LENGTH OF PIPE (feet)

DESIGN WATER FLOW (gpm)

feet	9	10	11	12	13	14	15	16	17	18	19	20	21	22	23	24	25	26	27	28	29	30	31	32
10	0	0	0	0	0	0	0	0	0	0	0	0	0	0	0	0	0	0	0	0	0	0	0	0
20	0	0	0	0	0	0	0	0	0	0	0	0	0	0	0	0	0	1	1	1	1	1	1	1
30	0	0	0	0	0	0	0	0	0	0	0	0	0	1	1	0	1	1	1	1	1	1	1	1
40	0	0	0	0	0	0	0	0	0	0	0	0	1	1	1	1	1	1	1	1	1	1	1	2
50	0	0	0	0	0	0	0	0	0	1	0	1	1	1	1	1	1	1	1	1	2	2	2	2
60	0	0	0	0	0	0	0	1	1	1	1	1	1	1	1	1	1	2	2	2	2	2	2	2
70	0	0	0	0	1	0	1	1	1	1	1	1	1	1	1	1	2	2	2	2	2	2	3	3
80	0	0	0	1	1	1	1	1	1	1	1	1	1	1	2	2	2	2	2	2	3	3	3	3
90	0	0	1	1	1	1	1	1	1	1	1	1	1	2	2	2	2	2	3	3	3	3	3	3
100	0	0	1	1	1	1	1	1	1	1	1	2	2	2	2	2	2	3	3	3	3	3	4	4
110	0	0	1	1	1	1	1	1	1	1	1	2	2	2	2	3	3	3	3	3	3	4	4	4
120	0	1	1	1	1	1	1	1	1	2	2	2	2	2	2	3	3	3	3	4	4	4	4	5
130	0	1	1	1	1	1	1	1	2	2	2	2	2	3	3	3	3	4	4	4	4	4	5	5
140	1	1	1	1	1	1	1	2	2	2	2	2	3	3	3	3	4	4	4	4	4	5	5	5
150	1	1	1	1	1	1	2	2	2	2	2	3	3	3	3	4	4	4	4	4	5	5	5	6
160	1	1	1	1	1	2	2	2	2	2	3	3	3	3	4	4	4	4	5	5	5	5	6	6
170	1	1	1	1	2	2	2	2	2	2	3	3	4	4	4	4	4	5	5	5	5	6	6	7
180	1	1	1	1	2	2	2	2	3	3	3	3	4	4	4	4	5	5	5	5	6	6	7	7
190	1	1	1	2	2	2	2	2	3	3	3	3	4	4	4	4	5	5	5	6	6	6	7	7
200	1	1	1	2	2	2	2	2	3	3	3	3	4	4	4	4	5	5	6	6	6	7	7	8

C = 150   i.d. = 1.481"

PRESSURE LOSS TABLE (psi)

2" COPPER PIPE (any type)

LENGTH OF PIPE ft \\ DESIGN WATER FLOW (gpm)	9	10	11	12	13	14	15	16	17	18	19	20	21	22	23	24	25	26	27	28	29	30	31	32
10	0	0	0	0	0	0	0	0	0	0	0	0	0	0	0	0	0	0	0	0	0	0	0	0
20	0	0	0	0	0	0	0	0	0	0	0	0	0	0	0	0	0	0	0	0	0	0	0	0
30	0	0	0	0	0	0	0	0	0	0	0	0	0	0	0	0	0	0	0	0	0	0	0	0
40	0	0	0	0	0	0	0	0	0	0	0	0	0	0	0	0	0	0	0	0	0	0	0	0
50	0	0	0	0	0	0	0	0	0	0	0	0	0	0	0	0	0	0	0	0	0	0	0	0
60	0	0	0	0	0	0	0	0	0	0	0	0	0	0	0	0	0	0	0	0	0	0	0	1
70	0	0	0	0	0	0	0	0	0	0	0	0	0	0	0	0	0	1	1	1	1	1	1	1
80	0	0	0	0	0	0	0	0	0	0	0	0	0	0	0	0	1	1	1	1	1	1	1	1
90	0	0	0	0	0	0	0	0	0	0	0	0	0	0	1	1	1	1	1	1	1	1	1	1
100	0	0	0	0	0	0	0	0	0	0	0	0	0	1	1	1	1	1	1	1	1	1	1	1
110	0	0	0	0	0	0	0	0	0	0	0	1	1	1	1	1	1	1	1	1	1	1	1	1
120	0	0	0	0	0	0	0	0	0	0	1	1	1	1	1	1	1	1	1	1	1	1	1	1
130	0	0	0	0	0	0	0	0	0	1	1	1	1	1	1	1	1	1	1	1	1	1	1	1
140	0	0	0	0	0	0	0	0	0	1	1	1	1	1	1	1	1	1	1	1	1	1	1	1
150	0	0	0	0	0	0	0	0	1	1	1	1	1	1	1	1	1	1	1	1	1	1	1	1
160	0	0	0	0	0	0	0	1	1	1	1	1	1	1	1	1	1	1	1	1	1	1	1	2
170	0	0	0	0	0	0	1	1	1	1	1	1	1	1	1	1	1	1	1	1	1	2	2	2
180	0	0	0	0	0	0	1	1	1	1	1	1	1	1	1	1	1	1	1	1	2	2	2	2
190	0	0	0	0	0	0	0	1	1	1	1	1	1	1	1	1	1	1	1	2	2	2	2	2
200	0	0	0	0	0	0	0	1	1	1	1	1	1	1	1	1	1	1	2	2	2	2	2	2

C = 150   i.d. = 1.959"

PRESSURE LOSS TABLE (psi)

1" POLYBUTELENE (PB) TUBING - SDR 9

LENGTH OF PIPE (ft)	DESIGN WATER FLOW (gpm)																							
	9	10	11	12	13	14	15	16	17	18	19	20	21	22	23	24	25	26	27	28	29	30	31	32
10	0	1	1	1	1	1	1	1	2	2	2	2	2	2	3	3	3	3	4	4	4	4	5	5
20	1	1	1	2	2	2	2	3	3	3	4	4	5	5	5	6	6	7	7	8	8	9	9	10
30	1	2	2	2	3	3	4	4	5	5	6	6	7	7	8	9	9	10	11	12	12	13	14	15
40	2	2	3	3	4	4	5	6	6	7	8	8	9	10	11	12	13	14	15	16	17	18	19	20
50	2	3	3	4	5	5	6	7	8	9	9	10	11	12	13	15	16	17	18	19	21	22	23	25
60	3	3	4	5	6	6	7	8	9	10	11	12	14	15	16	18	19	20	22	23	25	26	28	30
70	3	4	5	6	7	8	9	10	11	12	13	15	16	17	19	20	22	24	25	27	29	31	33	35
80	4	5	6	6	7	9	10	11	12	14	15	17	18	20	22	23	25	27	29	31	33	35	37	-
90	4	5	6	7	8	10	11	12	14	15	17	19	20	22	24	26	28	30	33	35	37	-	-	-
100	5	6	7	8	9	11	12	14	15	17	19	21	23	25	27	29	31	34	36	39	-	-	-	-
110	5	6	8	9	10	12	13	15	17	19	21	23	25	27	30	32	35	37	-	-	-	-	-	-
120	6	7	8	10	11	13	15	17	18	21	23	25	27	30	32	35	38	-	-	-	-	-	-	-
130	6	7	9	11	12	14	16	18	20	22	25	27	30	32	35	38	-	-	-	-	-	-	-	-
140	7	8	10	11	13	15	17	19	22	24	26	29	32	35	38	-	-	-	-	-	-	-	-	-
150	7	9	10	12	14	16	18	21	23	26	28	31	34	37	-	-	-	-	-	-	-	-	-	-
160	8	9	11	13	15	17	20	22	25	27	30	33	36	-	-	-	-	-	-	-	-	-	-	-
170	8	10	12	14	16	18	21	23	26	29	32	35	39	-	-	-	-	-	-	-	-	-	-	-
180	9	10	12	15	17	19	22	25	28	31	34	37	-	-	-	-	-	-	-	-	-	-	-	-
190	9	11	13	15	18	20	23	26	29	33	36	-	-	-	-	-	-	-	-	-	-	-	-	-
200	10	12	14	16	19	21	24	28	31	34	38	-	-	-	-	-	-	-	-	-	-	-	-	-

C = 150   i.d. = 0.875"

PRESSURE LOSS TABLE (psi)

1 1/4" POLYBUTELENE (PB) TUBING - SDR 9

C = 150   i.d. = 1.069"

LENGTH OF PIPE (ft)	DESIGN WATER FLOW (gpm)																							
	9	10	11	12	13	14	15	16	17	18	19	20	21	22	23	24	25	26	27	28	29	30	31	32
10	0	0	0	0	0	0	0	1	1	1	1	1	1	1	1	1	1	1	1	1	2	2	2	2
20	0	0	1	1	1	1	1	1	1	1	1	2	2	2	2	2	2	3	3	3	3	3	4	4
30	1	1	1	1	1	1	1	2	2	2	2	2	3	3	3	3	4	4	4	4	5	5	5	6
40	1	1	1	1	1	2	2	2	2	3	3	3	3	4	4	4	5	5	5	6	6	7	7	7
50	1	1	1	2	2	2	2	3	3	3	4	4	4	5	5	6	6	6	7	7	8	8	9	9
60	1	1	2	2	2	2	3	3	3	4	4	5	5	6	6	7	7	8	8	9	9	10	11	11
70	1	2	2	2	2	3	3	4	4	5	5	5	6	7	7	8	8	9	10	10	11	12	12	13
80	1	2	2	2	3	3	4	4	5	5	6	6	7	7	8	9	9	10	11	12	12	13	14	15
90	2	2	2	3	3	4	4	5	5	6	6	7	8	8	9	10	11	11	12	13	14	15	16	17
100	2	2	3	3	4	4	5	5	6	6	7	8	9	9	10	11	12	13	14	15	16	17	18	19
110	2	2	3	3	4	4	5	6	6	7	8	9	9	10	11	12	13	14	15	16	17	18	19	21
120	2	3	3	4	4	5	6	6	7	8	9	9	10	11	12	13	14	15	16	18	19	20	21	22
130	2	3	3	4	5	5	6	7	8	8	9	10	11	12	13	14	15	17	18	19	20	22	23	24
140	3	3	4	4	5	6	6	7	8	9	10	11	12	13	14	15	17	18	19	20	22	23	25	26
150	3	3	4	5	5	6	7	8	9	10	11	12	13	14	15	17	18	19	21	22	23	25	27	28
160	3	3	4	5	6	6	7	8	9	10	11	13	14	15	16	18	19	20	22	23	25	27	28	30
170	3	4	4	5	6	7	8	9	10	11	12	13	15	16	17	19	20	22	23	25	27	28	30	32
180	3	4	5	5	6	7	8	9	10	12	13	14	15	17	18	20	21	23	25	26	28	30	32	34
190	3	4	5	6	7	8	9	10	11	12	14	15	16	18	19	21	23	24	26	28	30	32	34	36
200	4	4	5	6	7	8	9	10	12	13	14	16	17	19	20	22	24	26	27	29	31	33	35	37

PRESSURE LOSS TABLE (psi)

1 1/2" POLYBUTELENE (PB) TUBING - SDR 9

DESIGN WATER FLOW (gpm)

LENGTH OF PIPE f t	9	10	11	12	13	14	15	16	17	18	19	20	21	22	23	24	25	26	27	28	29	30	31	32
10	0	0	0	0	0	0	0	0	0	0	0	0	0	0	0	0	1	1	1	1	1	1	1	1
20	0	0	0	0	0	0	0	0	1	1	1	1	1	1	1	1	1	1	2	1	1	1	2	2
30	0	0	0	0	0	0	0	0	1	1	1	1	2	2	2	1	2	2	2	2	2	2	2	2
40	0	0	0	1	1	1	1	1	1	1	1	1	2	2	2	2	2	2	2	3	3	3	3	3
50	0	0	0	1	1	1	1	1	1	2	2	2	2	2	3	3	3	3	3	3	3	4	4	4
60	0	1	1	1	1	1	1	2	2	2	2	2	2	3	3	3	3	3	4	4	4	4	5	5
70	1	1	1	1	1	1	2	2	2	2	2	3	3	3	4	4	4	4	4	5	5	5	5	6
80	1	1	1	1	2	2	2	2	2	3	3	3	3	4	4	4	4	5	5	5	6	6	6	7
90	1	1	1	1	2	2	2	2	3	3	3	4	4	4	5	5	5	5	6	6	6	7	7	7
100	1	1	1	2	2	2	2	3	3	3	4	4	4	5	5	5	6	6	6	7	7	7	8	8
110	1	1	1	2	2	2	3	3	3	4	4	4	5	5	6	6	6	7	7	8	8	8	9	9
120	1	1	2	2	2	2	3	3	3	4	4	5	5	6	6	6	7	7	8	8	9	9	9	10
130	1	1	2	2	2	3	3	3	4	4	5	5	6	6	7	7	7	8	8	9	9	10	10	11
140	1	1	2	2	2	3	3	4	4	5	5	6	6	7	7	8	8	9	9	10	10	11	11	12
150	1	2	2	2	3	3	3	4	4	5	5	6	6	7	8	8	8	9	9	10	11	11	12	12
160	1	2	2	2	3	3	4	4	5	5	6	6	7	8	8	9	9	10	10	11	11	12	13	13
170	1	2	2	3	3	3	4	4	5	5	6	7	7	8	9	9	10	10	11	11	12	13	13	14
180	1	2	2	3	3	4	4	5	5	6	6	7	8	8	9	10	10	11	12	12	13	13	14	15
190	2	2	2	3	3	4	4	5	5	6	7	7	8	9	9	10	11	11	12	13	13	14	15	16
200	2	2	2	3	4	4	4	5	5	6	7	7	8	8	9	10	11	11	12	13	14	15	16	17

C = 150   i.d. = 1.263"

PRESSURE LOSS TABLE (psi)

2" POLYBUTELENE (PB) TUBING - SDR 9

C = 150   i.d. = 1.653"

LENGTH OF PIPE (ft)	DESIGN WATER FLOW (gpm)																							
	9	10	11	12	13	14	15	16	17	18	19	20	21	22	23	24	25	26	27	28	29	30	31	32
10	0	0	0	0	0	0	0	0	0	0	0	0	0	0	0	0	0	0	0	0	0	0	0	0
20	0	0	0	0	0	0	0	0	0	0	0	0	0	0	0	0	0	0	0	0	0	0	0	0
30	0	0	0	0	0	0	0	0	0	0	0	0	0	0	1	1	1	1	1	1	1	1	1	1
40	0	0	0	0	0	0	0	0	0	0	0	0	0	0	1	1	1	1	1	1	1	1	1	1
50	0	0	0	0	0	0	0	0	0	0	0	0	0	1	1	1	1	1	1	1	1	1	1	1
60	0	0	0	0	0	0	0	0	0	0	0	0	1	1	1	1	1	1	1	1	1	1	1	1
70	0	0	0	0	0	0	0	0	0	0	1	1	1	1	1	1	1	1	1	1	1	1	1	2
80	0	0	0	0	0	0	0	0	1	1	1	1	1	1	1	1	1	1	1	2	2	2	2	2
90	0	0	0	0	0	0	0	1	1	1	1	1	1	1	1	1	1	2	2	2	2	2	2	2
100	0	0	0	0	0	0	1	1	1	1	1	1	1	1	1	2	2	2	2	2	2	2	2	2
110	0	0	0	0	0	1	1	1	1	1	1	1	1	1	2	2	2	2	2	2	2	2	2	2
120	0	0	0	0	1	1	1	1	1	1	1	1	1	1	2	2	2	2	2	2	2	2	3	3
130	0	0	0	0	1	1	1	1	1	1	1	1	2	2	2	2	2	2	2	2	3	3	3	3
140	0	0	0	0	1	1	1	1	1	1	1	1	2	2	2	2	2	2	2	3	3	3	3	3
150	0	0	0	0	1	1	1	1	1	1	1	2	2	2	2	2	2	2	3	3	3	3	3	3
160	0	0	0	0	1	1	1	1	1	1	1	2	2	2	2	2	2	3	3	3	3	3	3	4
170	0	0	0	0	1	1	1	1	1	1	2	2	2	2	2	3	3	3	3	3	3	3	4	4
180	0	0	1	1	1	1	1	1	1	1	2	2	2	2	2	3	3	3	3	3	3	4	4	4
190	0	0	1	1	1	1	1	1	1	1	2	2	2	2	2	3	3	3	3	4	4	4	4	4
200	0	1	1	1	1	1	1	1	1	2	2	2	2	2	2	3	3	3	3	4	4	4	4	4

PRESSURE LOSS TABLE (psi)

1" POLYETHYLENE (PE) PIPE - SIDR-PR

LENGTH OF PIPE (ft)	DESIGN WATER FLOW (gpm)																							
	9	10	11	12	13	14	15	16	17	18	19	20	21	22	23	24	25	26	27	28	29	30	31	32
10	0	0	0	0	0	0	1	1	1	1	1	1	1	1	1	1	1	1	2	2	2	2	2	2
20	0	0	1	1	1	1	1	1	1	1	2	2	2	2	2	2	3	3	3	3	3	4	4	4
30	1	1	1	1	1	1	2	2	2	2	2	3	3	3	3	4	4	4	5	5	5	5	6	6
40	1	1	1	1	2	2	2	2	3	3	3	3	4	4	4	5	5	6	6	6	7	7	8	8
50	1	1	1	2	2	2	3	3	3	4	4	4	5	5	5	6	7	7	8	8	9	9	10	10
60	1	1	2	2	2	3	3	3	4	4	5	5	6	6	6	7	8	8	9	9	10	11	12	12
70	1	2	2	2	3	3	4	4	4	5	5	6	7	7	7	8	9	10	11	11	12	13	14	14
80	2	2	2	3	3	4	4	5	5	6	6	7	8	8	8	10	10	11	12	13	14	15	15	16
90	2	2	3	3	3	4	5	5	6	6	7	8	8	9	9	11	12	13	14	14	15	16	17	18
100	2	2	3	3	4	4	5	6	6	7	8	9	9	10	10	12	13	14	15	16	17	18	19	21
110	2	3	3	4	4	5	6	6	7	8	9	9	10	11	11	13	14	15	17	18	19	20	21	23
120	2	3	4	4	5	5	6	7	8	9	9	10	11	12	12	14	16	17	18	19	21	22	23	25
130	3	3	4	4	5	6	7	7	8	9	10	11	12	13	13	16	17	18	20	21	22	24	25	27
140	3	3	4	5	5	6	7	8	9	10	11	12	13	14	14	17	18	20	21	22	24	26	27	29
150	3	4	5	5	6	7	8	8	10	11	12	13	14	15	16	18	20	21	23	24	26	27	29	31
160	3	4	5	5	6	7	8	9	10	11	13	14	15	16	17	19	21	22	24	26	27	29	31	33
170	3	4	5	6	7	8	9	10	11	12	13	15	16	17	18	21	22	24	26	27	29	31	33	35
180	4	4	5	6	7	8	9	10	11	13	14	16	17	18	19	22	23	25	27	29	31	33	35	37
190	4	5	5	6	7	8	10	11	12	13	15	16	18	20	20	23	25	27	29	31	33	35	37	-
200	4	5	6	7	8	9	10	11	13	14	16	17	19	21	22	24	26	28	30	32	34	36	39	-

C = 150   i.d. = 1.049"

1 1/4" POLYETHYLENE (PE) PIPE - SIDR-PR

DESIGN WATER FLOW (gpm)

C = 150   i.d. = 1.380"

LENGTH OF PIPE	9	10	11	12	13	14	15	16	17	18	19	20	21	22	23	24	25	26	27	28	29	30	31	32
10	0	0	0	0	0	0	0	0	0	0	0	0	0	0	0	0	0	0	0	0	0	0	1	1
20	0	0	0	0	0	0	0	0	0	0	0	0	0	1	1	1	1	1	1	1	1	1	1	1
30	0	0	0	0	0	0	0	0	0	1	1	1	1	1	1	1	1	1	1	1	1	1	2	2
40	0	0	0	0	0	0	0	1	1	1	1	1	1	1	1	1	1	1	1	1	1	1	2	2
50	0	0	0	0	1	1	1	1	1	1	1	1	1	1	1	1	1	2	2	2	2	2	3	3
60	0	0	0	1	1	1	1	1	1	1	1	1	1	1	2	2	2	2	2	2	2	3	3	3
70	0	0	1	1	1	1	1	1	1	1	1	1	2	2	2	2	2	3	3	3	3	3	4	4
80	0	1	1	1	1	1	1	1	1	1	2	2	2	2	2	2	3	3	3	3	4	4	4	4
90	0	1	1	1	1	1	1	1	2	2	2	2	2	2	2	3	3	3	3	4	4	4	5	5
100	1	1	1	1	1	1	1	2	2	2	2	2	2	3	3	3	3	3	4	4	4	4	5	5
110	1	1	1	1	1	1	2	2	2	2	2	3	3	3	3	3	4	4	4	4	5	5	6	6
120	1	1	1	1	1	2	2	2	2	2	3	3	3	3	4	4	4	4	5	5	5	5	6	6
130	1	1	1	1	2	2	2	2	2	3	3	3	3	4	4	4	5	5	5	5	6	6	7	7
140	1	1	1	2	2	2	2	2	3	3	3	3	4	4	4	5	5	6	6	6	6	7	7	8
150	1	1	2	2	2	2	2	3	3	3	3	4	4	4	5	5	5	6	6	6	7	7	8	8
160	1	1	2	2	2	2	3	3	3	3	4	4	4	5	5	5	6	6	6	7	7	7	8	9
170	1	1	2	2	2	2	3	3	3	4	4	4	5	5	5	6	6	6	7	7	8	8	9	9
180	1	1	2	2	2	3	3	3	4	4	4	4	5	5	6	6	7	7	7	8	8	9	9	10
190	1	1	2	2	3	3	3	3	4	4	4	5	5	6	6	6	7	7	8	8	9	9	10	10
200	1	2	2	2	3	3	3	3	4	4	5	5	5	6	6	7	7	8	8	8	9	10	10	11

PRESSURE LOSS TABLE (psi)

1 1/2" POLYETHYLENE (PE) PIPE - SIDR-PR

C = 150   i.d. = 1.610"

LENGTH OF PIPE (feet)	DESIGN WATER FLOW (gpm)																							
	9	10	11	12	13	14	15	16	17	18	19	20	21	22	23	24	25	26	27	28	29	30	31	32
10	0	0	0	0	0	0	0	0	0	0	0	0	0	0	0	0	0	0	0	0	0	0	0	0
20	0	0	0	0	0	0	0	0	0	0	0	0	0	0	0	0	0	0	0	0	0	0	0	1
30	0	0	0	0	0	0	0	0	0	0	0	0	0	0	0	0	0	1	1	1	1	1	1	1
40	0	0	0	0	0	0	0	0	0	0	0	0	0	1	1	1	1	1	1	1	1	1	1	1
50	0	0	0	0	0	0	0	0	0	0	0	1	1	1	1	1	1	1	1	1	1	1	1	1
60	0	0	0	0	0	0	0	0	0	1	1	1	1	1	1	1	1	1	1	1	1	1	1	2
70	0	0	0	0	0	0	0	0	1	1	1	1	1	1	1	1	1	1	1	1	2	2	2	2
80	0	0	0	0	0	0	1	1	1	1	1	1	1	1	1	1	1	1	2	2	2	2	2	2
90	0	0	0	0	0	1	1	1	1	1	1	1	1	1	1	1	1	2	2	2	2	2	2	2
100	0	0	0	0	0	1	1	1	1	1	1	1	1	1	1	2	2	2	2	2	2	2	2	3
110	0	0	0	0	1	1	1	1	1	1	1	1	1	1	2	2	2	2	2	2	2	3	3	3
120	0	0	0	1	1	1	1	1	1	1	1	1	1	2	2	2	2	2	2	2	3	3	3	3
130	0	0	0	1	1	1	1	1	1	1	1	1	2	2	2	2	2	2	2	3	3	3	3	3
140	0	0	0	1	1	1	1	1	1	1	1	2	2	2	2	2	2	2	3	3	3	3	3	4
150	0	0	1	1	1	1	1	1	1	1	1	2	2	2	2	2	2	3	3	3	3	3	4	4
160	0	0	1	1	1	1	1	1	1	1	2	2	2	2	2	2	3	3	3	3	3	4	4	4
170	0	1	1	1	1	1	1	1	1	2	2	2	2	2	2	3	3	3	3	3	4	4	4	4
180	0	1	1	1	1	1	1	1	1	2	2	2	2	2	3	3	3	3	3	4	4	4	4	5
190	0	1	1	1	1	1	1	1	2	2	2	2	2	2	3	3	3	3	4	4	4	4	5	5
200	0	1	1	1	1	1	1	1	2	2	2	2	2	3	3	3	3	4	4	4	4	5	5	5

PRESSURE LOSS TABLE (psi)

2" POLYETHYLENE (PE) PIPE - SIDR-PR

DESIGN WATER FLOW (gpm)

LENGTH OF PIPE ft	9	10	11	12	13	14	15	16	17	18	19	20	21	22	23	24	25	26	27	28	29	30	31	32
10	0	0	0	0	0	0	0	0	0	0	0	0	0	0	0	0	0	0	0	0	0	0	0	0
20	0	0	0	0	0	0	0	0	0	0	0	0	0	0	0	0	0	0	0	0	0	0	0	0
30	0	0	0	0	0	0	0	0	0	0	0	0	0	0	0	0	0	0	0	0	0	0	0	0
40	0	0	0	0	0	0	0	0	0	0	0	0	0	0	0	0	0	0	0	0	0	0	0	0
50	0	0	0	0	0	0	0	0	0	0	0	0	0	0	0	0	0	0	0	0	0	0	0	0
60	0	0	0	0	0	0	0	0	0	0	0	0	0	0	0	0	0	0	0	0	0	0	0	0
70	0	0	0	0	0	0	0	0	0	0	0	0	0	0	0	0	0	0	0	0	0	0	0	1
80	0	0	0	0	0	0	0	0	0	0	0	0	0	0	0	0	0	0	0	0	0	1	1	1
90	0	0	0	0	0	0	0	0	0	0	0	0	0	0	0	0	0	0	1	1	1	1	1	1
100	0	0	0	0	0	0	0	0	0	0	0	0	0	0	0	0	1	1	1	1	1	1	1	1
110	0	0	0	0	0	0	0	0	0	0	0	0	0	0	0	1	1	1	1	1	1	1	1	1
120	0	0	0	0	0	0	0	0	0	0	0	0	0	0	1	1	1	1	1	1	1	1	1	1
130	0	0	0	0	0	0	0	0	0	0	0	0	0	1	1	1	1	1	1	1	1	1	1	1
140	0	0	0	0	0	0	0	0	0	0	0	0	1	1	1	1	1	1	1	1	1	1	1	1
150	0	0	0	0	0	0	0	0	0	0	0	1	1	1	1	1	1	1	1	1	1	1	1	1
160	0	0	0	0	0	0	0	0	0	0	0	1	1	1	1	1	1	1	1	1	1	1	1	1
170	0	0	0	0	0	0	0	0	0	0	1	1	1	1	1	1	1	1	1	1	1	1	1	1
180	0	0	0	0	0	0	0	0	0	1	1	1	1	1	1	1	1	1	1	1	1	1	1	1
190	0	0	0	0	0	0	0	0	0	1	1	1	1	1	1	1	1	1	1	1	1	1	1	1
200	0	0	0	0	0	0	0	0	0	1	1	1	1	1	1	1	1	1	1	1	1	1	1	2

C = 150   i.d. = 2.067"

PRESSURE LOSS TABLE (psi)

3/4" CPVC PIPE - SDR-13.5

LENGTH OF PIPE (ft)	DESIGN WATER FLOW (gpm)																							
	9	10	11	12	13	14	15	16	17	18	19	20	21	22	23	24	25	26	27	28	29	30	31	32
10	0	1	1	1	1	1	1	1	1	2	2	2	2	2	3	3	3	3	3	4	4	4	4	5
20	1	1	1	2	2	2	2	3	3	3	4	4	4	5	5	6	6	6	7	7	8	8	9	9
30	1	2	2	2	3	3	3	4	4	5	5	6	7	7	8	8	9	10	10	11	12	13	13	14
40	2	2	3	3	4	4	5	5	6	7	7	8	9	9	10	11	12	13	14	15	16	17	18	19
50	2	3	3	4	4	5	6	7	7	8	9	10	11	12	13	14	15	16	18	18	20	21	22	24
60	3	3	4	5	5	6	7	8	9	10	11	12	13	14	16	17	18	19	21	22	24	25	27	28
70	3	4	5	5	6	7	8	9	10	11	13	14	15	17	18	19	21	23	24	26	28	29	31	33
80	4	4	5	6	7	8	9	10	12	13	14	16	17	19	21	22	24	26	28	30	32	34	36	38
90	4	5	6	7	8	9	10	12	13	15	16	18	20	21	23	25	27	29	31	33	35	38	-	-
100	5	5	7	8	9	10	12	13	15	16	18	20	22	24	26	28	30	32	35	37	-	-	-	-
110	5	6	7	8	10	11	13	14	16	18	20	22	24	26	28	31	33	35	38	-	-	-	-	-
120	5	7	8	9	11	12	14	16	18	20	22	24	26	28	31	33	36	39	-	-	-	-	-	-
130	6	7	9	10	12	13	15	17	19	21	23	26	28	31	33	36	39	-	-	-	-	-	-	-
140	6	8	9	11	13	14	16	18	21	23	25	28	30	33	36	39	-	-	-	-	-	-	-	-
150	7	8	10	12	13	15	17	20	22	24	27	30	33	35	38	-	-	-	-	-	-	-	-	-
160	7	9	10	12	14	16	19	21	23	26	29	32	35	38	-	-	-	-	-	-	-	-	-	-
170	8	9	11	13	15	17	20	22	25	28	31	34	37	-	-	-	-	-	-	-	-	-	-	-
180	8	10	12	14	16	18	21	24	26	29	32	36	-	-	-	-	-	-	-	-	-	-	-	-
190	9	10	12	15	17	19	22	25	28	31	34	38	-	-	-	-	-	-	-	-	-	-	-	-
200	9	11	13	15	18	20	23	26	29	33	36	-	-	-	-	-	-	-	-	-	-	-	-	-

C = 150   i.d. = 0.884"

PRESSURE LOSS TABLE (psi)

1" CPVC PIPE - SDR-13.5

DESIGN WATER FLOW (gpm)

LENGTH OF PIPE Ft	9	10	11	12	13	14	15	16	17	18	19	20	21	22	23	24	25	26	27	28	29	30	31	32
10	0	0	0	0	0	0	0	0	0	1	1	1	1	1	1	1	1	1	1	1	1	1	1	2
20	0	0	0	1	1	1	1	1	1	1	1	1	2	2	2	2	2	2	2	2	3	3	3	3
30	0	1	1	1	1	1	1	2	2	2	2	2	2	3	3	3	3	3	3	4	4	4	4	5
40	1	1	1	1	1	1	2	2	2	3	3	3	3	3	3	4	4	4	5	5	5	6	6	6
50	1	1	1	1	2	2	2	2	3	3	3	4	4	4	4	5	5	5	6	6	7	7	7	8
60	1	1	1	2	2	2	3	3	3	4	4	4	5	5	5	6	6	6	7	7	8	8	9	9
70	1	1	2	2	2	2	3	3	4	4	5	5	6	6	6	6	7	7	8	9	9	10	10	11
80	1	1	2	2	3	3	3	3	4	5	5	6	6	7	7	7	8	9	9	10	10	11	12	13
90	1	2	2	3	3	3	4	4	4	5	5	6	6	7	8	8	9	10	10	11	12	13	13	14
100	1	2	2	3	3	4	4	4	5	6	6	7	7	8	9	9	10	11	11	12	13	14	15	16
110	2	2	2	3	3	4	4	5	5	6	7	7	8	9	9	10	11	12	13	13	14	15	16	17
120	2	2	3	4	4	4	5	5	6	6	7	8	9	9	10	11	12	13	14	15	16	17	18	19
130	2	2	3	4	4	5	5	6	6	7	8	9	10	10	11	12	13	14	15	16	17	18	19	20
140	2	3	3	4	5	5	6	6	7	8	9	9	10	11	12	13	14	15	16	17	18	19	21	22
150	2	3	4	4	5	6	6	7	7	8	9	10	11	12	13	14	15	16	17	18	20	21	22	24
160	2	3	4	5	5	6	6	7	8	9	10	11	12	13	14	15	16	17	18	20	21	22	24	25
170	3	3	4	5	6	6	7	8	9	10	10	12	13	14	15	16	17	18	19	21	22	24	25	27
180	3	3	4	5	6	7	7	8	9	10	11	12	13	14	15	16	18	19	21	22	24	25	27	28
190	3	3	5	6	6	7	8	8	9	11	11	13	14	15	16	17	19	20	22	23	25	26	28	30
200	3	4	5	6	6	7	8	9	10	11	12	13	14	16	17	18	20	21	23	24	26	28	30	31

C = 150   i.d. = 1.109"

PRESSURE LOSS TABLE (psi)

1 1/4" CPVC PIPE - SDR-13.5

C = 150   i.d. = 1.400"

DESIGN WATER FLOW (gpm)

LENGTH OF PIPE	9	10	11	12	13	14	15	16	17	18	19	20	21	22	23	24	25	26	27	28	29	30	31	32
10	0	0	0	0	0	0	0	0	0	0	0	0	0	0	0	0	0	0	0	0	0	0	1	1
20	0	0	0	0	0	0	0	0	0	0	0	1	0	1	1	1	1	1	1	1	1	1	1	1
30	0	0	0	0	0	0	0	0	1	1	1	1	1	1	1	1	1	1	1	1	1	1	1	2
40	0	0	0	0	0	0	0	1	1	1	1	1	1	1	1	1	1	2	2	2	2	2	2	2
50	0	0	0	0	0	0	1	1	1	1	1	1	1	2	2	2	2	2	2	2	2	2	2	3
60	0	0	0	0	1	1	1	1	1	1	2	2	2	2	2	2	2	2	2	3	3	3	3	3
70	0	0	0	1	1	1	1	1	1	1	2	2	2	2	2	2	3	3	3	3	3	3	3	4
80	0	0	1	1	1	1	1	1	2	2	2	2	2	2	3	3	3	3	3	3	4	4	4	4
90	0	1	1	1	1	1	1	1	2	2	2	2	2	3	3	3	3	4	4	4	4	4	4	5
100	0	1	1	1	1	1	2	2	2	2	2	2	3	3	3	3	3	4	4	4	4	4	5	5
110	1	1	1	1	1	2	2	2	2	2	2	3	3	3	3	4	4	4	5	5	5	5	5	6
120	1	1	1	1	2	2	2	2	2	3	3	3	3	3	4	4	4	5	5	5	5	5	6	6
130	1	1	1	1	2	2	2	2	2	3	3	3	3	4	4	4	5	5	5	5	6	6	6	7
140	1	1	1	2	2	2	2	2	3	3	3	3	4	4	4	4	5	5	6	6	6	6	7	7
150	1	1	1	2	2	2	2	3	3	3	3	4	4	4	5	5	5	6	6	6	6	7	7	8
160	1	1	1	2	2	2	2	3	3	3	4	4	4	5	5	5	6	6	6	7	7	7	8	8
170	1	1	2	2	2	3	3	3	3	3	4	4	5	5	5	6	6	6	7	7	7	8	8	9
180	1	1	2	2	2	3	3	3	3	4	4	4	5	5	5	6	6	7	7	7	8	8	9	9
190	1	1	2	2	2	3	3	3	3	4	4	5	5	5	6	6	6	7	7	8	8	8	9	10
200	1	1	2	2	2	3	3	3	3	4	4	5	5	5	6	6	7	7	8	8	8	9	10	10

PRESSURE LOSS TABLE (psi)

1 1/2" CPVC PIPE - SDR-13.5

DESIGN WATER FLOW (gpm)

LENGTH OF PIPE	9	10	11	12	13	14	15	16	17	18	19	20	21	22	23	24	25	26	27	28	29	30	31	32
10	0	0	0	0	0	0	0	0	0	0	0	0	0	0	0	0	0	0	0	0	0	0	0	0
20	0	0	0	0	0	0	0	0	0	0	0	0	0	0	0	0	0	0	0	0	0	0	0	1
30	0	0	0	0	0	0	0	0	0	0	0	0	0	0	0	0	0	1	1	1	1	1	1	1
40	0	0	0	0	0	0	0	0	0	0	0	0	0	1	1	1	1	1	1	1	1	1	1	1
50	0	0	0	0	0	0	0	0	0	0	0	1	1	1	1	1	1	1	1	1	1	1	1	1
60	0	0	0	0	0	0	0	0	0	1	1	1	1	1	1	1	1	1	1	1	1	1	1	2
70	0	0	0	0	0	0	0	1	1	1	1	1	1	1	1	1	1	1	1	1	2	2	2	2
80	0	0	0	0	0	0	1	1	1	1	1	1	1	1	1	1	1	1	2	2	2	2	2	2
90	0	0	0	0	0	1	1	1	1	1	1	1	1	1	1	1	1	2	2	2	2	2	2	2
100	0	0	0	0	0	1	1	1	1	1	1	1	1	1	1	2	2	2	2	2	2	2	2	3
110	0	0	0	0	1	1	1	1	1	1	1	1	1	1	2	2	2	2	2	2	2	3	3	3
120	0	0	0	1	1	1	1	1	1	1	1	1	1	2	2	2	2	2	2	2	3	3	3	3
130	0	0	0	1	1	1	1	1	1	1	1	1	2	2	2	2	2	2	2	3	3	3	3	3
140	0	0	1	1	1	1	1	1	1	1	1	2	2	2	2	2	2	2	3	3	3	3	3	4
150	0	0	1	1	1	1	1	1	1	1	1	2	2	2	2	2	2	3	3	3	3	3	4	4
160	0	0	1	1	1	1	1	1	1	1	2	2	2	2	2	2	3	3	3	3	3	4	4	4
170	0	1	1	1	1	1	1	1	1	2	2	2	2	2	2	3	3	3	3	3	4	4	4	4
180	0	1	1	1	1	1	1	1	1	2	2	2	2	2	3	3	3	3	3	4	4	4	4	5
190	0	1	1	1	1	1	1	1	2	2	2	2	2	2	3	3	3	3	4	4	4	4	5	5
200	1	1	1	1	1	1	1	1	2	2	2	2	2	3	3	3	3	4	4	4	4	5	5	5

C = 150   i.d. = 1.602"

PRESSURE LOSS TABLE (psi)

2" CPVC PIPE - SDR-13.5

C = 150   i.d. = 2.003"

DESIGN WATER FLOW (gpm)

LENGTH OF PIPE (ft)	9	10	11	12	13	14	15	16	17	18	19	20	21	22	23	24	25	26	27	28	29	30	31	32
10	0	0	0	0	0	0	0	0	0	0	0	0	0	0	0	0	0	0	0	0	0	0	0	0
20	0	0	0	0	0	0	0	0	0	0	0	0	0	0	0	0	0	0	0	0	0	0	0	0
30	0	0	0	0	0	0	0	0	0	0	0	0	0	0	0	0	0	0	0	0	0	0	0	0
40	0	0	0	0	0	0	0	0	0	0	0	0	0	0	0	0	0	0	0	0	0	0	0	0
50	0	0	0	0	0	0	0	0	0	0	0	0	0	0	0	0	0	0	0	0	0	0	0	0
60	0	0	0	0	0	0	0	0	0	0	0	0	0	0	0	0	0	0	0	0	0	0	1	1
70	0	0	0	0	0	0	0	0	0	0	0	0	0	0	0	0	0	0	0	0	1	1	1	1
80	0	0	0	0	0	0	0	0	0	0	0	0	0	0	0	0	0	0	1	1	1	1	1	1
90	0	0	0	0	0	0	0	0	0	0	0	0	0	0	0	0	1	1	1	1	1	1	1	1
100	0	0	0	0	0	0	0	0	0	0	0	0	0	0	0	1	1	1	1	1	1	1	1	1
110	0	0	0	0	0	0	0	0	0	0	0	0	0	0	1	1	1	1	1	1	1	1	1	1
120	0	0	0	0	0	0	0	0	0	0	0	0	0	1	1	1	1	1	1	1	1	1	1	1
130	0	0	0	0	0	0	0	0	0	0	0	0	1	1	1	1	1	1	1	1	1	1	1	1
140	0	0	0	0	0	0	0	0	0	0	0	1	1	1	1	1	1	1	1	1	1	1	1	1
150	0	0	0	0	0	0	0	0	0	0	1	1	1	1	1	1	1	1	1	1	1	1	1	1
160	0	0	0	0	0	0	0	0	0	0	1	1	1	1	1	1	1	1	1	1	1	1	1	1
170	0	0	0	0	0	0	0	0	0	1	1	1	1	1	1	1	1	1	1	1	1	1	1	2
180	0	0	0	0	0	0	0	0	0	1	1	1	1	1	1	1	1	1	1	1	1	1	2	2
190	0	0	0	0	0	0	0	0	1	1	1	1	1	1	1	1	1	1	1	1	1	1	2	2
200	0	0	0	0	0	0	0	0	1	1	1	1	1	1	1	1	1	1	1	1	1	2	2	2

PRESSURE LOSS TABLE (psi)

1" PVC PIPE - SDR-21

LENGTH OF PIPE (ft)	DESIGN WATER FLOW (gpm) 9	10	11	12	13	14	15	16	17	18	19	20	21	22	23	24	25	26	27	28	29	30	31	32	LENGTH OF PIPE (ft)
10	0	0	0	0	0	0	0	0	0	0	0	0	1	1	1	1	1	1	1	1	1	1	1	1	10
20	0	0	0	0	0	0	1	1	1	1	1	1	1	1	1	1	1	2	2	2	2	2	2	2	20
30	0	0	0	1	1	1	1	1	1	1	1	1	2	2	2	2	2	2	2	3	3	3	3	3	30
40	0	1	1	1	1	1	1	1	1	2	2	2	2	2	2	3	3	3	3	3	4	4	4	4	40
50	1	1	1	1	1	1	1	2	2	2	2	2	3	3	3	3	4	4	4	4	5	5	5	6	50
60	1	1	1	1	1	1	2	2	2	2	3	3	3	3	4	4	4	5	5	5	6	6	6	7	60
70	1	1	1	1	1	2	2	2	2	3	3	3	4	4	4	5	5	5	6	6	7	7	7	8	70
80	1	1	1	1	2	2	2	2	3	3	3	4	4	4	5	5	6	6	7	7	7	8	8	9	80
90	1	1	1	2	2	2	2	3	3	3	4	4	5	5	5	6	6	7	7	8	8	9	9	10	90
100	1	1	2	2	2	2	3	3	3	4	4	5	5	6	6	7	7	8	8	9	9	10	11	11	100
110	1	1	2	2	2	3	3	3	4	4	5	5	6	6	7	7	8	8	9	10	10	11	12	12	110
120	1	2	2	2	3	3	3	4	4	5	5	6	6	7	7	8	8	9	10	10	11	12	13	13	120
130	1	2	2	2	3	3	4	4	5	5	6	6	7	7	8	9	9	10	11	11	12	13	14	15	130
140	1	2	2	3	3	3	4	4	5	5	6	7	7	8	8	9	10	11	11	12	13	14	15	16	140
150	2	2	2	3	3	4	4	5	5	6	6	7	8	8	9	10	11	11	12	13	14	15	16	17	150
160	2	2	2	3	3	4	4	5	6	6	7	7	8	9	10	10	11	12	13	14	15	16	17	18	160
170	2	2	3	3	4	4	5	5	6	7	7	8	9	9	10	11	12	13	14	15	16	17	18	19	170
180	2	2	3	3	4	4	5	6	6	7	8	8	9	10	11	12	13	14	15	16	17	18	19	20	180
190	2	2	3	3	4	5	5	6	7	7	8	9	10	11	12	12	13	14	15	17	18	19	20	21	190
200	2	3	3	4	4	5	5	6	7	8	9	9	10	11	12	13	14	15	16	17	19	20	21	22	200

C = 150 i.d.= 1.189"

PRESSURE LOSS TABLE (psi)

1 1/4" PVC PIPE - SDR-21

	DESIGN WATER FLOW (gpm)																							
LENGTH OF PIPE (feet)	9	10	11	12	13	14	15	16	17	18	19	20	21	22	23	24	25	26	27	28	29	30	31	32
10	0	0	0	0	0	0	0	0	0	0	0	0	0	0	0	0	0	0	0	0	0	0	0	0
20	0	0	0	0	0	0	0	0	0	0	0	0	0	0	0	0	0	0	1	1	1	1	1	1
30	0	0	0	0	0	0	0	0	0	0	0	0	0	1	1	1	1	1	1	1	1	1	1	1
40	0	0	0	0	0	0	0	0	0	0	1	1	1	1	1	1	1	1	1	1	1	1	1	1
50	0	0	0	0	0	0	0	0	1	1	1	1	1	1	1	1	1	1	1	1	1	2	2	2
60	0	0	0	0	0	0	1	1	1	1	1	1	1	1	1	1	1	1	2	2	2	2	2	2
70	0	0	0	0	0	1	1	1	1	1	1	1	1	1	1	1	2	2	2	2	2	2	2	2
80	0	0	0	0	1	1	1	1	1	1	1	1	1	1	2	2	2	2	2	2	2	3	3	3
90	0	0	0	1	1	1	1	1	1	1	1	1	1	2	2	2	2	2	2	3	3	3	3	3
100	0	0	0	1	1	1	1	1	1	1	1	1	2	2	2	2	2	2	3	3	3	3	3	4
110	0	0	1	1	1	1	1	1	1	1	1	2	2	2	2	2	2	3	3	3	3	3	4	4
120	0	0	1	1	1	1	1	1	1	1	2	2	2	2	2	3	3	3	3	3	4	4	4	4
130	0	1	1	1	1	1	1	1	1	2	2	2	2	2	3	3	3	3	3	4	4	4	4	5
140	0	1	1	1	1	1	1	1	2	2	2	2	2	2	3	3	3	3	4	4	4	4	5	5
150	1	1	1	1	1	1	1	1	2	2	2	2	2	3	3	3	3	4	4	4	4	5	5	5
160	1	1	1	1	1	1	1	2	2	2	2	2	3	3	3	3	4	4	4	4	5	5	5	6
170	1	1	1	1	1	1	1	2	2	2	2	3	3	3	3	4	4	4	4	5	5	5	6	6
180	1	1	1	1	1	1	2	2	2	2	2	3	3	3	3	4	4	4	5	5	5	6	6	6
190	1	1	1	1	1	1	2	2	2	2	3	3	3	3	4	4	4	5	5	5	6	6	6	7
200	1	1	1	1	1	2	2	2	2	2	3	3	3	4	4	4	5	5	5	6	6	6	7	7

C = 150   i.d. = 1.502"

PRESSURE LOSS TABLE (psi)

1 1/2" PVC PIPE - SDR-21

LENGTH OF PIPE (Ft)	DESIGN WATER FLOW (gpm)																							
	9	10	11	12	13	14	15	16	17	18	19	20	21	22	23	24	25	26	27	28	29	30	31	32
10	0	0	0	0	0	0	0	0	0	0	0	0	0	0	0	0	0	0	0	0	0	0	0	0
20	0	0	0	0	0	0	0	0	0	0	0	0	0	0	0	0	0	0	0	0	0	0	0	0
30	0	0	0	0	0	0	0	0	0	0	0	0	0	0	0	0	0	0	0	0	0	0	1	1
40	0	0	0	0	0	0	0	0	0	0	0	0	0	0	0	0	0	1	1	1	1	1	1	1
50	0	0	0	0	0	0	0	0	0	0	0	0	0	0	0	1	1	1	1	1	1	1	1	1
60	0	0	0	0	0	0	0	0	0	0	0	0	1	1	1	1	1	1	1	1	1	1	1	1
70	0	0	0	0	0	0	0	0	0	0	0	1	1	1	1	1	1	1	1	1	1	1	1	1
80	0	0	0	0	0	0	0	0	0	1	1	1	1	1	1	1	1	1	1	1	1	1	1	1
90	0	0	0	0	0	0	0	0	1	1	1	1	1	1	1	1	1	1	1	1	1	1	2	2
100	0	0	0	0	0	0	0	1	1	1	1	1	1	1	1	1	1	1	1	1	2	2	2	2
110	0	0	0	0	0	0	0	1	1	1	1	1	1	1	1	1	1	1	1	2	2	2	2	2
120	0	0	0	0	0	0	1	1	1	1	1	1	1	1	1	1	1	2	2	2	2	2	2	2
130	0	0	0	0	1	1	1	1	1	1	1	1	1	1	1	1	2	2	2	2	2	2	2	2
140	0	0	0	0	1	1	1	1	1	1	1	1	1	1	1	2	2	2	2	2	2	2	2	3
150	0	0	0	0	1	1	1	1	1	1	1	1	1	1	1	2	2	2	2	2	2	2	3	3
160	0	0	0	0	1	1	1	1	1	1	1	1	1	1	2	2	2	2	2	2	2	3	3	3
170	0	0	0	1	1	1	1	1	1	1	1	1	1	2	2	2	2	2	2	2	3	3	3	3
180	0	0	0	1	1	1	1	1	1	1	1	1	2	2	2	2	2	2	2	3	3	3	3	3
190	0	0	0	1	1	1	1	1	1	1	1	1	2	2	2	2	2	2	3	3	3	3	3	4
200	0	0	1	1	1	1	1	1	1	1	1	2	2	2	2	2	2	3	3	3	3	3	3	4

C = 150   i.d. = 1.720"

PRESSURE LOSS TABLE (psi)

2" PVC PIPE - SDR-21

C = 150 i.d. = 2.149"

LENGTH OF PIPE (feet)	DESIGN WATER FLOW (gpm)																								
	9	10	11	12	13	14	15	16	17	18	19	20	21	22	23	24	25	26	27	28	29	30	31	32	
10	0	0	0	0	0	0	0	0	0	0	0	0	0	0	0	0	0	0	0	0	0	0	0	0	10
20	0	0	0	0	0	0	0	0	0	0	0	0	0	0	0	0	0	0	0	0	0	0	0	0	20
30	0	0	0	0	0	0	0	0	0	0	0	0	0	0	0	0	0	0	0	0	0	0	0	0	30
40	0	0	0	0	0	0	0	0	0	0	0	0	0	0	0	0	0	0	0	0	0	0	0	0	40
50	0	0	0	0	0	0	0	0	0	0	0	0	0	0	0	0	0	0	0	0	0	0	0	0	50
60	0	0	0	0	0	0	0	0	0	0	0	0	0	0	0	0	0	0	0	0	0	0	0	0	60
70	0	0	0	0	0	0	0	0	0	0	0	0	0	0	0	0	0	0	0	0	0	0	0	0	70
80	0	0	0	0	0	0	0	0	0	0	0	0	0	0	0	0	0	0	0	0	0	0	0	0	80
90	0	0	0	0	0	0	0	0	0	0	0	0	0	0	0	0	0	0	0	0	0	0	1	1	90
100	0	0	0	0	0	0	0	0	0	0	0	0	0	0	0	0	0	0	0	0	1	1	1	1	100
110	0	0	0	0	0	0	0	0	0	0	0	0	0	0	0	0	0	0	1	1	1	1	1	1	110
120	0	0	0	0	0	0	0	0	0	0	0	0	0	0	0	0	0	1	1	1	1	1	1	1	120
130	0	0	0	0	0	0	0	0	0	0	0	0	0	0	0	0	1	1	1	1	1	1	1	1	130
140	0	0	0	0	0	0	0	0	0	0	0	0	0	0	0	1	1	1	1	1	1	1	1	1	140
150	0	0	0	0	0	0	0	0	0	0	0	0	0	0	1	1	1	1	1	1	1	1	1	1	150
160	0	0	0	0	0	0	0	0	0	0	0	0	0	0	1	1	1	1	1	1	1	1	1	1	160
170	0	0	0	0	0	0	0	0	0	0	0	0	0	1	1	1	1	1	1	1	1	1	1	1	170
180	0	0	0	0	0	0	0	0	0	0	0	0	1	1	1	1	1	1	1	1	1	1	1	1	180
190	0	0	0	0	0	0	0	0	0	0	0	0	1	1	1	1	1	1	1	1	1	1	1	1	190
200	0	0	0	0	0	0	0	0	0	0	0	1	1	1	1	1	1	1	1	1	1	1	1	1	200

APPENDIX B
TABLES FOR
MAXIMUM SPRINKLER STRAIGHT PIPE RUNS

ALLOWABLE INSIDE PIPE LENGTHS AT 10 GPM DESIGN WATER FLOW (DWF)

CHOOSE 1 ROW USING - ONE COLUMN FROM PIPE SECTION A PLUS ONE COLUMN FROM PIPE SECTION B

10 GPM	INSIDE PIPE SECTION A		INSIDE PIPE SECTION B						
	CPVC OR CU (M) 1 1/4"	CPVC OR CU (M) 1"	CU (M) 1"	CU (M) 3/4"	CPVC 1"	CPVC 3/4"	CPVC 3/4 S"	PB 1"	PB 3/4"
AVAILABLE PRESSURE FOR PIPING									
10	-	-	191	53	244	81	28	99	29
10	25	-	182	51	232	77	27	94	27
10	50	-	173	48	220	73	26	89	26
10	75	-	163	45	208	69	24	84	25
10	-	25	166	46	212	70	25	86	25
10	-	50	141	39	180	60	21	73	21
10	-	75	116	32	148	49	17	60	17
15	-	-	287	80	366	121	43	148	43
15	25	-	278	77	354	117	41	143	42
15	50	-	268	75	342	113	40	138	40
15	75	-	259	72	330	109	38	134	39
15	-	25	262	73	334	111	39	135	39
15	-	50	237	66	302	100	35	122	36
15	-	75	212	59	270	90	31	109	32
20	-	-	383	106	488	162	57	197	58
20	25	-	373	104	476	158	55	193	56
20	50	-	364	101	464	154	54	188	55
20	75	-	355	98	452	150	53	183	53
20	-	25	358	99	456	151	53	185	54
20	-	50	333	92	424	141	49	172	50
20	-	75	308	85	392	130	46	159	46
25	-	-	478	133	610	202	71	247	72
25	25	-	469	130	598	198	70	242	71
25	50	-	460	128	586	194	68	237	69
25	75	-	450	125	574	190	67	232	68
25	-	25	453	126	578	192	67	234	68
25	-	50	428	119	546	181	64	221	64
25	-	75	403	112	514	170	60	208	61
30	-	-	574	159	732	243	85	296	86
30	25	-	565	157	720	239	84	291	85
30	50	-	555	154	708	235	82	286	83
30	75	-	546	152	696	231	81	282	82
30	-	25	549	152	700	232	81	283	83
30	-	50	524	146	668	221	78	270	79
30	-	75	499	139	636	211	74	257	75
35	-	-	670	186	854	283	99	346	101
35	25	-	660	183	842	279	98	341	99
35	50	-	651	181	830	275	97	336	98
35	75	-	641	178	818	271	95	331	96
35	-	25	645	179	822	272	96	333	97
35	-	50	620	172	790	262	92	320	93
35	-	75	595	165	758	251	88	307	89

Table numbers represent 44% of the total pipe length available at the given pressure (fittings included @ 56% of total pipe length).

To find the total pipe length available at the given pressure, multiply the column value times 2.25.

ALLOWABLE INSIDE PIPE LENGTHS AT 12 GPM DESIGN WATER FLOW (DWF)

CHOOSE 1 ROW USING - ONE COLUMN FROM PIPE SECTION A PLUS ONE COLUMN FROM PIPE SECTION B

12 GPM	INSIDE PIPE SECTION A		INSIDE PIPE SECTION B						
AVAILABLE PRESSURE FOR PIPING	CPVC OR CU (M) 1 1/4"	CPVC OR CU (M) 1"	CU (M) 1"	CU (M) 3/4"	CPVC 1"	CPVC 3/4"	CPVC 3/4 S"	PB 1"	PB 3/4"
10	-	-	137	38	174	58	20	70	21
10	25	-	127	35	162	54	19	66	19
10	50	-	118	33	150	50	17	61	18
10	75	-	108	30	138	46	16	56	16
10	-	25	112	31	142	47	17	58	17
10	-	50	87	24	110	37	13	45	13
10	-	75	62	17	78	26	9	32	9
15	-	-	205	57	261	87	30	106	31
15	25	-	195	54	249	83	29	101	29
15	50	-	186	52	237	79	28	96	28
15	75	-	177	49	225	75	26	91	27
15	-	25	180	50	229	76	27	93	27
15	-	50	155	43	197	65	23	80	23
15	-	75	130	36	165	55	19	67	20
20	-	-	273	76	348	115	41	141	41
20	25	-	264	73	336	111	39	136	40
20	50	-	254	71	324	107	38	131	38
20	75	-	245	68	312	104	36	126	37
20	-	25	248	69	316	105	37	128	37
20	-	50	223	62	284	94	33	115	34
20	-	75	198	55	253	84	29	102	30
25	-	-	341	95	435	144	51	176	51
25	25	-	332	92	423	140	49	171	50
25	50	-	323	90	411	136	48	166	49
25	75	-	313	87	399	132	46	162	47
25	-	25	316	88	403	134	47	163	48
25	-	50	291	81	371	123	43	150	44
25	-	75	266	74	340	113	40	137	40
30	-	-	410	114	522	173	61	211	62
30	25	-	400	111	510	169	59	207	60
30	50	-	391	109	498	165	58	202	59
30	75	-	382	106	486	161	57	197	57
30	-	25	385	107	490	163	57	198	58
30	-	50	360	100	459	152	53	186	54
30	-	75	335	93	427	141	50	173	50
35	-	-	478	133	609	202	71	247	72
35	25	-	468	130	597	198	70	242	70
35	50	-	459	128	585	194	68	237	69
35	75	-	450	125	574	190	67	232	68
35	-	25	453	126	577	191	67	234	68
35	-	50	428	119	546	181	63	221	64
35	-	75	403	112	514	170	60	208	61

Table numbers represent 44% of the total pipe length available at the given pressure (fittings included @ 56% of total pipe length). To find he total pipe length available at the given pressure, multiply the column value times 225.

ALLOWABLE INSIDE PIPE LENGTHS AT 14 GPM DESIGN WATER FLOW (DWF)

CHOOSE 1 ROW USING - ONE COLUMN FROM PIPE SECTION A PLUS ONE COLUMN FROM PIPE SECTION B

14 GPM	INSIDE PIPE SECTION A		INSIDE PIPE SECTION B						
	CPVC OR CU (M) 1 1/4"	1"	CU (M) 1"	CU (M) 3/4"	CPVC 1"	CPVC 3/4"	CPVC 3/4 S"	PB 1"	PB 3/4"
10	-	-	103	29	131	43	15	53	15
10	25	-	93	26	119	39	14	48	14
10	50	-	84	23	107	35	12	43	13
10	75	-	75	21	95	32	11	38	11
10	-	25	78	22	99	33	12	40	12
10	-	50	53	15	67	22	8	27	8
10	-	75	28	8	35	12	4	14	4
15	-	-	154	43	196	65	23	79	23
15	25	-	145	40	184	61	21	75	22
15	50	-	135	38	172	57	20	70	20
15	75	-	126	35	161	53	19	65	19
15	-	25	129	36	164	55	19	67	19
15	-	50	104	29	133	44	15	54	16
15	-	75	79	22	101	33	12	41	12
20	-	-	205	57	262	87	30	106	31
20	25	-	196	54	250	83	29	101	29
20	50	-	187	52	238	79	28	96	28
20	75	-	177	49	226	75	26	91	27
20	-	25	180	50	230	76	27	93	27
20	-	50	155	43	198	66	23	80	23
20	-	75	130	36	166	55	19	67	20
25	-	-	257	71	327	108	38	132	39
25	25	-	247	69	315	105	37	128	37
25	50	-	238	66	303	101	35	123	36
25	75	-	229	63	291	97	34	118	34
25	-	25	232	64	295	98	34	120	35
25	-	50	207	57	263	87	31	107	31
25	-	75	182	50	232	77	27	94	27
30	-	-	308	86	393	130	46	159	46
30	25	-	299	83	381	126	44	154	45
30	50	-	289	80	369	122	43	149	43
30	75	-	280	78	357	118	42	144	42
30	-	25	283	79	361	120	42	146	43
30	-	50	258	72	329	109	38	133	39
30	-	75	233	65	297	98	35	120	35
35	-	-	359	100	458	152	53	185	54
35	25	-	350	97	446	148	52	181	53
35	50	-	341	95	434	144	51	176	51
35	75	-	331	92	422	140	49	171	50
35	-	25	334	93	426	141	50	173	50
35	-	50	309	86	394	131	46	160	47
35	-	75	284	79	362	120	42	147	43

(Left margin vertical label: A V A I L A B L E P R E S S U R E F O R P I P P I N G)

Table numbers represent 44% of the total pipe length available at the given pressure (fittings included @ 56% of total pipe length).
To find the total pipe length available at the given pressure, multiply the column value times 2.25.

ALLOWABLE INSIDE PIPE LENGTHS AT 16 GPM DESIGN WATER FLOW (DWF)

CHOOSE 1 ROW USING . ONE COLUMN FROM PIPE SECTION A PLUS ONE COLUMN FROM PIPE SECTION B

16 GPM	INSIDE PIPE SECTION A		INSIDE PIPE SECTION B						
AVAILABLE PRESSURE FOR PIPING	CPVC OR CU (M) 1 1/4"	CPVC OR CU (M) 1"	CU (M) 1"	CU (M) 3/4"	CPVC 1"	CPVC 3/4"	CPVC 3/4 S"	PB 1"	PB 3/4"
10	-	-	80	22	102	34	12	41	12
10	25	-	71	20	90	30	11	37	11
10	50	-	61	17	78	26	9	32	9
10	75	-	52	14	66	22	8	27	8
10	-	25	55	15	70	23	8	28	8
10	-	50	30	8	38	13	4	16	5
10	-	75	5	1	7	2	1	3	1
15	-	-	120	33	153	51	18	62	18
15	25	-	111	31	141	47	16	57	17
15	50	-	102	28	130	43	15	52	15
15	75	-	92	26	118	39	14	48	14
15	-	25	95	26	121	40	14	49	14
15	-	50	70	20	90	30	10	36	11
15	-	75	45	13	58	19	7	23	7
20	-	-	160	45	204	68	24	83	24
20	25	-	151	42	193	64	22	78	23
20	50	-	142	39	181	60	21	73	21
20	75	-	132	37	169	56	20	68	20
20	-	25	135	38	173	57	20	70	20
20	-	50	110	31	141	47	16	57	17
20	-	75	85	24	109	36	13	44	13
25	-	-	200	56	256	85	30	103	30
25	25	-	191	53	244	81	28	99	29
25	50	-	182	50	232	77	27	94	27
25	75	-	172	48	220	73	26	89	26
25	-	25	175	49	224	74	26	91	26
25	-	50	150	42	192	64	22	78	23
25	-	75	125	35	160	53	19	65	19
30	-	-	241	67	307	102	36	124	36
30	25	-	231	64	295	98	34	119	35
30	50	-	222	62	283	94	33	114	33
30	75	-	212	59	271	90	32	110	32
30	-	25	216	60	275	91	32	111	32
30	-	50	191	53	243	81	28	98	29
30	-	75	166	46	211	70	25	85	25
35	-	-	281	78	358	119	42	145	42
35	25	-	271	75	346	115	40	140	41
35	50	-	262	73	334	111	39	135	39
35	75	-	253	70	322	107	37	130	38
35	-	25	256	71	326	108	38	132	38
35	-	50	231	64	294	97	34	119	35
35	-	75	206	57	262	87	31	106	31

Table numbers represent 44% of the total pipe length available at the given pressure (fittings included @ 56% of total pipe length). To find the total pipe length available at the pressure, multiply the column value times 2.25.

ALLOWABLE INSIDE PIPE LENGTHS AT 18 GPM DESIGN WATER FLOW (DWF)

CHOOSE 1 ROW USING - ONE COLUMN FROM PIPE SECTION A PLUS ONE COLUMN FROM PIPE SECTION B

18 GPM

	INSIDE PIPE SECTION A		INSIDE PIPE SECTION B						
	CPVC OR CU (M) 1 1/4"	1"	CU (M) 1"	CU (M) 3/4"	CPVC 1"	CPVC 3/4"	CPVC 3/4 S"	PB 1"	PB 3/4"
15	-	-	97	27	123	41	14	50	15
15	25	-	87	24	111	37	13	45	13
15	50	-	78	22	99	33	12	40	12
15	75	-	69	19	88	29	10	35	10
15	-	25	72	20	91	30	11	37	11
15	-	50	47	13	60	20	7	24	7
15	-	75	22	6	28	9	3	11	3
20	-	-	129	36	164	55	19	67	19
20	25	-	120	33	153	51	18	62	18
20	50	-	110	31	141	47	16	57	17
20	75	-	101	28	129	43	15	52	15
20	-	25	104	29	133	44	15	54	16
20	-	50	79	22	101	33	12	41	12
20	-	75	54	15	69	23	8	28	8
25	-	-	161	45	206	68	24	83	24
25	25	-	152	42	194	64	23	78	23
25	50	-	142	40	182	60	21	74	21
25	75	-	133	37	170	56	20	69	20
25	-	25	136	38	174	58	20	70	20
25	-	50	111	31	142	47	16	57	17
25	-	75	86	24	110	36	13	44	13
30	-	-	193	54	247	82	29	100	29
30	25	-	184	51	235	78	27	95	28
30	50	-	175	49	223	74	26	90	26
30	75	-	165	46	211	70	25	85	25
30	-	25	168	47	215	71	25	87	25
30	-	50	143	40	183	61	21	74	22
30	-	75	118	33	151	50	18	61	18
35	-	-	226	63	288	95	33	116	34
35	25	-	216	60	276	91	32	112	33
35	50	-	207	57	264	87	31	107	31
35	75	-	198	55	252	84	29	102	30
35	-	25	201	56	256	85	30	104	30
35	-	50	176	49	224	74	26	91	26
35	-	75	151	42	192	64	22	78	23
40	-	-	258	72	329	109	38	133	39
40	25	-	249	69	317	105	37	128	37
40	50	-	239	66	305	101	35	123	36
40	75	-	230	64	293	97	34	119	35
40	-	25	233	65	297	98	35	120	35
40	-	50	208	58	265	88	31	107	31
40	-	75	183	51	233	77	27	94	28

(Left margin vertical label: AVAILABLE PRESSURE FOR PIPING)

Table numbers represent 44% of the total pipe length avauable at the given pressure (fittings included @ 56% of total pipe length)
To find the total pipe length available at the given pressure, multiply the column value times 2.25.

ALLOWABLE INSIDE PIPE LENGTHS AT 20 GPM DESIGN WATER FLOW (DWF)

CHOOSE 1 ROW USING - ONE COLUMN FROM PIPE SECTION A PLUS ONE COLUMN FROM PIPE SECTION B

20 GPM	INSIDE PIPE SECTION A		INSIDE PIPE SECTION B						
	CPVC OR CU (M)								
AVAILABLE PRESSURE FOR PIPING	1 1/4"	1"	CU (M) 1"	CU (M) 3/4"	CPVC 1"	CPVC 3/4"	CPVC 3/4 S"	PB 1"	PB 3/4"
15	-	-	80	22	101	34	12	41	12
15	25	-	70	20	90	30	10	36	11
15	50	-	61	17	78	26	9	31	9
15	75	-	52	14	66	22	8	27	8
15	-	25	55	15	70	23	8	28	8
15	-	50	30	8	38	13	4	15	4
15	-	75	5	1	6	2	1	2	1
20	-	-	106	29	135	45	16	55	16
20	25	-	97	27	123	41	14	50	15
20	50	-	87	24	111	37	13	45	13
20	75	-	78	22	100	33	12	40	12
20	-	25	81	23	103	34	12	42	12
20	-	50	56	16	72	24	8	29	8
20	-	75	31	9	40	13	5	16	5
25	-	-	133	37	169	56	20	68	20
25	25	-	123	34	157	52	18	64	19
25	50	-	114	32	145	48	17	59	17
25	75	-	105	29	133	44	16	54	16
25	-	25	108	30	137	46	16	56	16
25	-	50	83	23	105	35	12	43	12
25	-	75	58	16	74	24	9	30	9
30	-	-	159	44	203	67	24	82	24
30	25	-	150	42	191	63	22	77	23
30	50	-	140	39	179	59	21	72	21
30	75	-	131	36	167	55	19	68	20
30	-	25	134	37	171	57	20	69	20
30	-	50	109	30	139	46	16	56	16
30	-	75	84	23	107	36	12	43	13
35	-	-	186	52	237	78	28	96	28
35	25	-	176	49	225	75	26	91	27
35	50	-	167	46	213	71	25	86	25
35	75	-	158	44	201	67	23	81	24
35	-	25	161	45	205	68	24	83	24
35	-	50	136	38	173	57	20	70	20
35	-	75	111	31	141	47	16	57	17
40	-	-	212	59	271	90	31	110	32
40	25	-	203	56	259	86	30	105	31
40	50	-	194	54	247	82	29	100	29
40	75	-	184	51	235	78	27	95	28
40	-	25	187	52	239	79	28	97	28
40	-	50	162	45	207	69	24	84	24
40	-	75	137	38	175	58	20	71	21

Table numbers represent 44% of the total pipe length available at the given pressure (fittings included @ 55% of total pipe length). To find the total pipe length available at the given pressure, multiply the column value times 225.

ALLOWABLE INSIDE PIPE LENGTHS AT 22 GPM DESIGN WATER FLOW (DWF)

CHOOSE 1 ROW USING - ONE COLUMN FROM PIPE SECTION A PLUS ONE COLUMN FROM PIPE SECTION B

22 GPM

	INSIDE PIPE SECTION A		INSIDE PIPE SECTION B						
	CPVC OR CU (M) 1 1/4"	1"	CU (M) 1"	CU (M) 3/4"	CPVC 1"	CPVC 3/4"	CPVC 3/4 S"	PB 1"	PB 3/4"
15	-	-	67	19	85	28	10	34	10
15	25	-	57	16	73	24	9	30	9
15	50	-	48	13	61	20	7	25	7
15	75	-	39	11	49	16	6	20	6
15	-	25	42	12	53	18	6	22	6
15	-	50	17	5	21	7	2	9	3
15	-	-	-	-	-	-	-	-	-
20	-	-	89	25	113	38	13	46	13
20	25	-	80	22	102	34	12	41	12
20	50	-	70	20	90	30	10	36	11
20	75	-	61	17	78	26	9	31	9
20	-	25	64	18	82	27	9	33	10
20	-	50	39	11	50	16	6	20	6
20	-	75	14	4	18	6	2	7	2
25	-	-	111	31	142	47	16	57	17
25	25	-	102	28	130	43	15	53	15
25	50	-	93	26	118	39	14	48	14
25	75	-	83	23	106	35	12	43	13
25	-	25	86	24	110	36	13	44	13
25	-	50	61	17	78	26	9	32	9
25	-	75	36	10	46	15	5	19	5
30	-	-	133	37	170	56	20	69	20
30	25	-	124	34	158	52	18	64	19
30	50	-	115	32	146	48	17	59	17
30	75	-	105	29	134	45	16	54	16
30	-	25	108	30	138	46	16	56	16
30	-	50	83	23	106	35	12	43	13
30	-	75	58	16	75	25	9	30	9
35	-	-	156	43	199	66	23	80	23
35	25	-	146	41	187	62	22	76	22
35	50	-	137	38	175	58	20	71	21
35	75	-	128	35	163	54	19	66	19
35	-	25	131	36	167	55	19	67	20
35	-	50	106	29	135	45	16	55	16
35	-	75	81	22	103	34	12	42	12
40	-	-	178	49	227	75	26	92	27
40	25	-	169	47	215	71	25	87	25
40	50	-	159	44	203	67	24	82	24
40	75	-	150	42	191	63	22	77	23
40	-	25	153	42	195	65	23	79	23
40	-	50	128	36	163	54	19	66	19
40	-	75	103	29	131	44	15	53	15

(Left vertical label: AVAILABLE PRESSURE FOR PIPING)

Table numbers represent 44% of the total pipe length available at the given pressure (fittings included @ 56% of total pipe length).
To find the total pipe length available at the given pressure, multiply the column value times 2.25.

ALLOWABLE INSIDE PIPE LENGTHS AT 24 GPM DESIGN WATER FLOW (DWF)

CHOOSE 1 ROW USING - ONE COLUMN FROM PIPE SECTION A PLUS ONE COLUMN FROM PIPE SECTION B

24 GPM	INSIDE PIPE SECTION A		INSIDE PIPE SECTION B						
	CPVC OR CU (M) 1 1/4"	1"	CU (M) 1"	CU (M) 3/4"	CPVC 1"	CPVC 3/4"	CPVC 3/4 S"	PB 1"	PB 3/4"
15	-	-	57	16	72	24	8	29	9
15	25	-	47	13	61	20	7	24	7
15	50	-	38	11	49	16	6	20	6
15	75	-	29	8	37	12	4	15	4
15	-	25	32	9	41	13	5	16	5
15	-	50	7	2	9	3	1	4	1
15	-	-	-	-	-	-	-	-	-
20	-	-	76	21	97	32	11	39	11
20	25	-	66	18	85	28	10	34	10
20	50	-	57	16	73	24	8	29	9
20	75	-	48	13	61	20	7	25	7
20	-	25	51	14	65	21	8	26	8
20	-	50	26	7	33	11	4	13	4
20	-	75	1	0	1	0	0	0	0
25	-	-	95	26	121	40	14	49	14
25	25	-	85	24	109	36	13	44	13
25	50	-	76	21	97	32	11	39	11
25	75	-	67	19	85	28	10	34	10
25	-	25	70	19	89	29	10	36	10
25	-	50	45	12	57	19	7	23	7
25	-	75	20	5	25	8	3	10	3
30	-	-	114	32	145	48	17	59	17
30	25	-	104	29	133	44	15	54	16
30	50	-	95	26	121	40	14	49	14
30	75	-	86	24	109	36	13	44	13
30	-	25	89	25	113	37	13	46	13
30	-	50	64	18	81	27	9	33	10
30	-	75	39	11	49	16	6	20	6
35	-	-	133	37	169	56	20	68	20
35	25	-	123	34	157	52	18	64	19
35	50	-	114	32	145	48	17	59	17
35	75	-	104	29	133	44	16	54	16
35	-	25	108	30	137	45	16	56	16
35	-	50	83	23	105	35	12	43	12
35	-	75	58	16	73	24	9	30	9
40	-	-	151	42	193	64	22	78	23
40	25	-	142	39	181	60	21	73	21
40	50	-	133	37	169	56	20	69	20
40	75	-	123	34	157	52	18	64	19
40	-	25	126	35	161	53	19	65	19
40	-	50	101	28	129	43	15	52	15
40	-	75	76	21	98	32	11	39	12

Table numbers represent 44% of the total pipe length available at the given pressure (fittings included @ 56% of total pipe length)
To find the total pipe length available at given pressure, multiply the column times 2.25.

ALLOWABLE INSIDE PIPE LENGTHS AT 26 GPM DESIGN WATER FLOW (DWF)

CHOOSE 1 ROW USING-ONE COLUMN FROM PIPE SECTION A PLUS ONE COLUMN FROM PIPE SECTION B

26 GPM	INSIDE PIPE SECTION A		INSIDE PIPE SECTION B						
	CPVC OR CU (M) 1 1/4"	1"	CU (M) 1"	CU (M) 3/4"	CPVC 1"	CPVC 3/4"	CPVC 3/4 S"	PB 1"	PB 3/4"
20	-	-	65	18	83	28	10	34	10
20	25	-	56	16	71	24	8	29	8
20	50	-	47	13	59	20	7	24	7
20	75	-	37	10	48	16	6	19	6
20	-	25	40	11	51	17	6	21	6
20	-	50	15	4	20	6	2	8	2
20	-	-	-	-	-	-	-	-	-
25	-	-	82	23	104	35	12	42	12
25	25	-	72	20	92	31	11	37	11
25	50	-	63	17	80	27	9	32	9
25	75	-	54	15	68	23	8	28	8
25	-	25	57	16	72	24	8	29	9
25	-	50	32	9	40	13	5	16	5
25	-	75	7	2	8	3	1	3	1
30	-	-	98	27	125	41	15	51	15
30	25	-	89	25	113	37	13	46	13
30	50	-	79	22	101	34	12	41	12
30	75	-	70	19	89	30	10	36	11
30	-	25	73	20	93	31	11	38	11
30	-	50	48	13	61	20	7	25	7
30	-	75	23	6	29	10	3	12	3
35	-	-	114	32	146	48	17	59	17
35	25	-	105	29	134	44	16	54	16
35	50	-	96	27	122	40	14	49	14
35	75	-	86	24	110	36	13	45	13
35	-	25	89	25	114	38	13	46	13
35	-	50	64	18	82	27	10	33	10
35	-	75	39	11	50	17	6	20	6
40	-	-	131	36	167	55	19	67	20
40	25	-	121	34	155	51	18	63	18
40	50	-	112	31	143	47	17	58	17
40	75	-	103	28	131	43	15	53	15
40	-	25	106	29	135	45	16	55	16
40	-	50	81	22	103	34	12	42	12
40	-	75	56	15	71	24	8	29	8
45	-	-	147	41	187	62	22	76	22
45	25	-	138	38	175	58	20	71	21
45	50	-	128	36	164	54	19	66	19
45	75	-	119	33	152	50	18	61	18
45	-	25	122	34	156	52	18	63	18
45	-	50	97	27	124	41	14	50	15
45	-	75	72	20	92	30	11	37	11

(Left vertical label: A V A I L A B L E   P R E S S U R E   F O R   P I P I N G)

Table numbers represent 44% of the total pipe length available at the given pressure (fittings included @ 56% of total length).
To find the total pipe length available at the given pressure, multiply the column value times 225.

ALLOWABLE INSIDE PIPE LENGTHS AT 28 GPM DESIGN WATER FLOW (DWF)

CHOOSE 1 ROW USING - ONE COLUMN FROM PIPE SECTION A PLUS ONE COLUMN FROM PIPE SECTION B

28 GPM	INSIDE PIPE SECTION A		INSIDE PIPE SECTION B						
	CPVC OR CU (M)		**CU (M)**	**CU (M)**	**CPVC**	**CPVC**	**CPVC**	**PB**	**PB**
	1 1/4"	**1"**	**1"**	**3/4"**	**1"**	**3/4"**	**3/4 S"**	**1"**	**3/4"**
20	-	-	57	16	73	24	8	29	9
20	25	-	48	13	61	20	7	25	7
20	50	-	38	11	49	16	6	20	6
20	75	-	29	8	37	12	4	15	4
20	-	25	32	9	41	14	5	16	5
20	-	50	7	2	9	3	1	4	1
20	-	-	-	-	-	-	-	-	-
25	-	-	71	20	91	30	11	37	11
25	25	-	62	17	79	26	9	32	9
25	50	-	52	15	67	22	8	27	8
25	75	-	43	12	55	18	6	22	6
25	-	25	46	13	59	20	7	24	7
25	-	50	21	6	27	9	3	11	3
25	-	-	-	-	-	-	-	-	-
30	-	-	85	24	109	36	13	44	13
30	25	-	76	21	97	32	11	39	11
30	50	-	67	19	85	28	10	34	10
30	75	-	57	16	73	24	9	30	9
30	-	25	60	17	77	26	9	31	9
30	-	50	35	10	45	15	5	18	5
30	-	75	10	3	13	4	2	5	2
35	-	-	100	28	127	42	15	51	15
35	25	-	90	25	115	38	13	47	14
35	50	-	81	22	103	34	12	42	12
35	75	-	72	20	91	30	11	37	11
35	-	25	75	21	95	32	11	39	11
35	-	50	50	14	63	21	7	26	7
35	-	75	25	7	31	10	4	13	4
40	-	-	114	32	145	48	17	59	17
40	25	-	105	29	133	44	16	54	16
40	50	-	95	26	121	40	14	49	14
40	75	-	86	24	109	36	13	44	13
40	-	25	89	25	113	38	13	46	13
40	-	50	64	18	81	27	9	33	10
40	-	75	39	11	50	16	6	20	6
45	-	-	128	36	163	54	19	66	19
45	25	-	119	33	151	50	18	61	18
45	50	-	109	30	140	46	16	56	16
45	75	-	100	28	128	42	15	52	15
45	-	25	103	29	132	44	15	53	16
45	-	50	78	22	100	33	12	40	12
45	-	75	53	15	68	22	8	27	8

(Left margin vertical label: AVAILABLE PRESSURE FOR PIPING)

Table numbers represent 44% of the total pipe length available at the given pressure (fittings included @ 56% of total pipe length)
To find the total pipe length available at the given pressure, multiply the column value times 2.25.

ALLOWABLE INSIDE PIPE LENGTHS AT 30 GPM DESIGN WATER FLOW (DWF)

CHOOSE 1 ROW USING - ONE COLUMN FROM PIPE SECTION A PLUS ONE COLUMN FROM PIPE SECTION B

A V A I L A B L E   P R E S S U R E   F O R   P I P I N G

30 GPM	INSIDE PIPE SECTION A CPVC OR CU (M) 1 1/4"	1"	CU (M) 1"	CU (M) 3/4"	CPVC 1"	CPVC 3/4"	CPVC 3/4 S"	PB 1"	PB 3/4"
20	-	-	50	14	64	21	7	26	8
20	25	-	41	11	52	17	6	21	6
20	50	-	31	9	40	13	5	16	5
20	75	-	22	6	28	9	3	11	3
20	-	25	25	7	32	11	4	13	4
20	-	50	0	0	0	0	0	0	0
20	-	-	-	-	-	-	-	-	-
25	-	-	63	17	80	26	9	32	9
25	25	-	53	15	68	23	8	28	8
25	50	-	44	12	56	19	7	23	7
25	75	-	35	10	44	15	5	18	5
25	-	25	38	10	48	16	6	19	6
25	-	50	13	4	16	5	2	7	2
25	-	-	-	-	-	-	-	-	-
30	-	-	75	21	96	32	11	39	11
30	25	-	66	18	84	28	10	34	10
30	50	-	56	16	72	24	8	29	8
30	75	-	47	13	60	20	7	24	7
30	-	25	50	14	64	21	7	26	8
30	-	50	25	7	32	11	4	13	4
30	-	75	0	0	0	0	0	0	0
35	-	-	88	24	112	37	13	45	13
35	25	-	78	22	100	33	12	40	12
35	50	-	69	19	88	29	10	36	10
35	75	-	60	17	76	25	9	31	9
35	-	25	63	17	80	27	9	32	9
35	-	50	38	10	48	16	6	19	6
35	-	75	13	4	16	5	2	7	2
40	-	-	100	28	128	42	15	52	15
40	25	-	91	25	116	38	13	47	14
40	50	-	82	23	104	34	12	42	12
40	75	-	72	20	92	31	11	37	11
40	-	25	75	21	96	32	11	39	11
40	-	50	50	14	64	21	7	26	8
40	-	75	25	7	32	11	4	13	4
45	-	-	113	31	144	48	17	58	17
45	25	-	103	29	132	44	15	53	16
45	50	-	94	26	120	40	14	49	14
45	75	-	85	24	108	36	13	44	13
45	-	25	88	24	112	37	13	45	13
45	-	50	63	17	80	27	9	32	9
45	-	75	38	10	48	16	6	19	6

Table numbers represent 44% of the total pipe length available at the given pressure (fittings included @ 56% of total pipe length).
To find the total available pipe length at the given pressure, multiply the column values times 2 25.

ALLOWABLE INSIDE PIPE LENGTHS AT 32 GPM DESIGN WATER FLOW (DWF)

CHOOSE 1 ROW USING - ONE COLUMN FROM PIPE SECTION A PLUS ONE COLUMN FROM PIPE SECTION B

32 GPM

AVAILABLE PRESSURE FOR PIPING

	INSIDE PIPE SECTION A		INSIDE PIPE SECTION B						
	CPVC OR CU (M) 1 1/4"	1"	CU (M) 1"	CU (M) 3/4"	CPVC 1"	CPVC 3/4"	CPVC 3/4 S"	PB 1"	PB 3/4"
20	-	-	44	12	57	19	7	23	7
20	25	-	35	10	45	15	5	18	5
20	50	-	26	7	33	11	4	13	4
20	75	-	16	5	21	7	2	8	2
20	-	25	19	5	25	8	3	10	3
20	-	-	-	-	-	-	-	-	-
20	-	-	-	-	-	-	-	-	-
25	-	-	56	15	71	24	8	29	8
25	25	-	46	13	59	20	7	24	7
25	50	-	37	10	47	16	5	19	6
25	75	-	28	8	35	12	4	14	4
25	-	25	31	9	39	13	5	16	5
25	-	50	6	2	7	2	1	3	1
25	-	-	-	-	-	-	-	-	-
30	-	-	67	19	85	28	10	34	10
30	25	-	57	16	73	24	9	30	9
30	50	-	48	13	61	20	7	25	7
30	75	-	39	11	49	16	6	20	6
30	-	25	42	12	53	18	6	22	6
30	-	50	17	5	21	7	2	9	3
30	-	-	-	-	-	-	-	-	-
35	-	-	78	22	99	33	12	40	12
35	25	-	68	19	87	29	10	35	10
35	50	-	59	16	75	25	9	31	9
35	75	-	50	14	63	21	7	26	7
35	-	25	53	15	67	22	8	27	8
35	-	50	28	8	36	12	4	14	4
35	-	75	3	1	4	1	0	1	0
40	-	-	89	25	113	38	13	46	13
40	25	-	80	22	102	34	12	41	12
40	50	-	70	20	90	30	10	36	11
40	75	-	61	17	78	26	9	31	9
40	-	25	64	18	82	27	9	33	10
40	-	50	39	11	50	16	6	20	6
40	-	75	14	4	18	6	2	7	2
45	-	-	100	28	128	42	15	52	15
45	25	-	91	25	116	38	13	47	14
45	50	-	81	23	104	34	12	42	12
45	75	-	72	20	92	30	11	37	11
45	-	25	75	21	96	32	11	39	11
45	-	50	50	14	64	21	7	26	8
45	-	75	25	7	32	11	4	13	4

Table numbers represent 44% of the total pipe length available at the given pressure (fittings included @ 56% of total pipe length).
To find the total available pipe length at the given pressure, multiply the column value times 2.25.

APPENDIX C

SPRINKLER LAYOUT TEMPLETS

DESIGN GUIDE WORKSHEET

RESIDENTIAL SPRINKLER

TARGET ZONE TEMPLET
1/4" = 1' scale

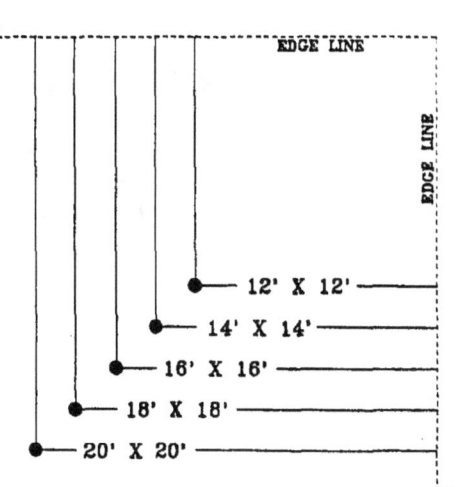

MINIMUM DISTANCE
BETWEEN SPRINKLERS

8'–0"

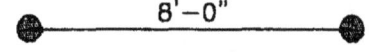

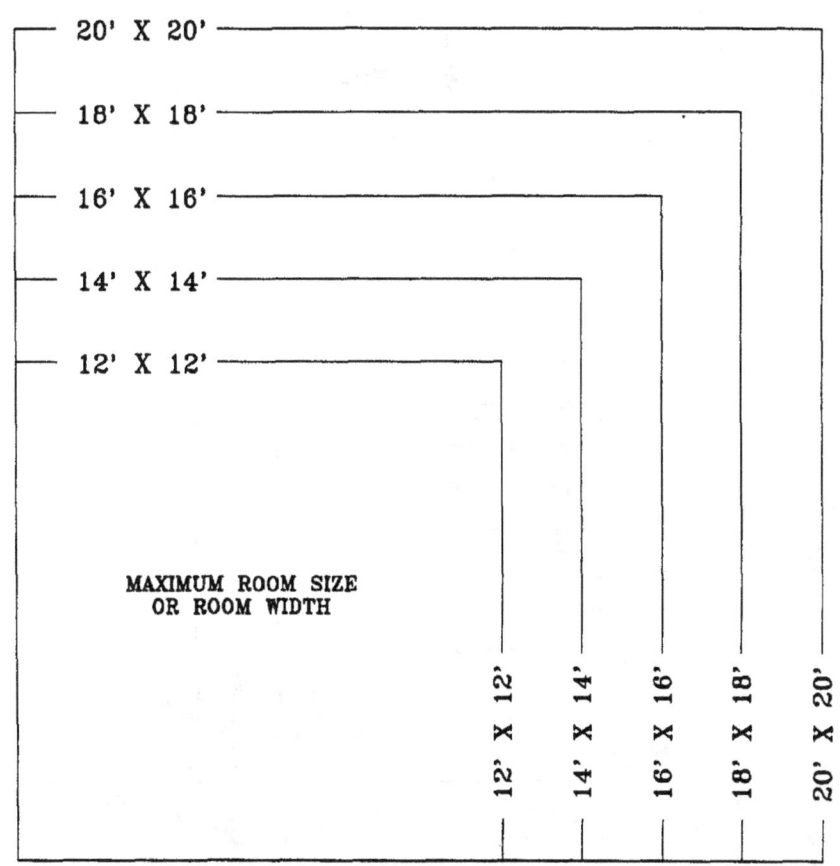

RESIDENTIAL SPRINKLER

TARGET ZONE TEMPLET
1/8" = 1' scale

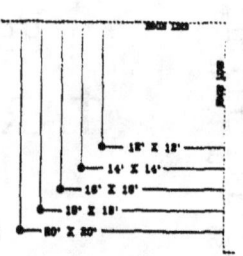

MINIMUM DISTANCE
BETWEEN SPRINKLERS

8'-0"

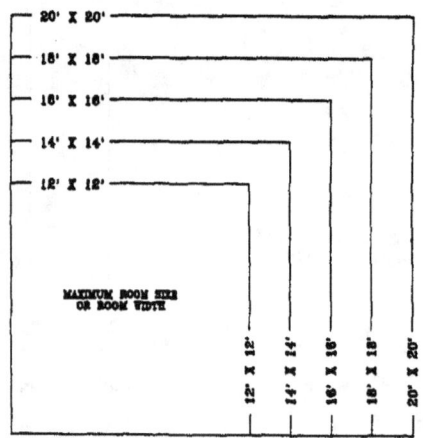

HYDRAULIC WORKSHEET

1. **ROOM WIDTH AND COVERAGE AREA**

   A. Room Width:_____ft.

   B. Coverage Area: _____ ft. x _____ ft.

2. **SPRINKLER HEAD SPECIFICATIONS**

   A. Single-Head Flow Rate: _____ gpm.

   B. Single-Head Pressure: _____ psi.

   C. Dual-Head Flow Rate:_____ gpm.

3. **DESIGN WATER FLOW (DWF) AND DESIGN PRESSURE**

   A. If all rooms have only one sprinkler head:

         DWF (from 2A):_____g p m.

   B. If more than one head in any room:

         DWF (From 2C): _____ x 2 = _____ gpm.

   Design Water Flow (A or B above):   **Line 1:_____ gpm**

   Design Sprinkler Pressure (From 2B):   **Line 2:_____ psi**

4. **WATER PRESSURE AT THE PUBLIC MAIN**   **Line 3: _____ psi**

5. **PRESSURE LOSSES CAUSED BY DEVICES**

   A. Backflow Prevention Device; Check Valve   **Line 4: - - - - - P s i**

   B. Water Meter Loss

         Water Meter Size: _____

         Pressure Loss
         (Use DWF on Line 1, and Table 3)   **Line 5:_____ psi**

   C. Gate or Ball Valve Loss
      (Use DWF and Table 3)
      _____ X _____ psi =   **Line 6:_____ psi**
      No. Valves   Loss

6. **PRESSURE LOSSES IN UNDERGROUND SUPPLY PIPING**

   Find the Pressure Losses based on the DWF on Line 1
   and Tables in Appendix A.

   A. Underground Section #1 Piping

   _____ , _____ , _____ ft: Pressure Loss =
   Size   Type   Length   Line 7A:_____ p s i

   B. Underground Section #2 Piping

   _____ , _____ , _____ ft: Pressure Loss =
   Sire   Type   Length   Line 7B:_____ p s i

7. **ELEVATION PRESSURE LOSS**

   Difference in elevation between water main tap point and
   highest sprinkler (if the sprinkler head is lower, the
   number is negative): _____ / 2 =   Line 8:_____ p s i

8. **SUM OF LOSSES AND SPRINKLER PRESSURE**

   _____ + _____ + _____ + _____ +
   Line 2   Line 4   Line 5   Line 6

   _____ + _____ + _____ =
   Line 7A + Line 7B + Line 8   Line 9:_____ p s i

9. **AVAILABLE PRESSURE FOR PIPING**

   _____ - _____ =
   Line 3   Line 9   Line 10:_____ psi

10. **SELECTION OF PIPE TYPE AND SIZE**

    Use the appropriate Table in Appendix B, based on the
    DWF, Line 1. Find the Available Pressure for Piping,
    Line 9, in the Table's left-hand column. Select the
    piping type(s) and size(s).

INSIDE SECTION A: _____ , _____ ,_____ ft. maximum straight length
                    Type   Size

INSIDE SECTION B: _____ , _____ ,_____ ft. maximum straight length
                    Type   Size

APPENDIX D

INSIDE DIAMETER TABLE
FITTING LOSS TABLE
FRICTION LOSS TABLE

INSIDE DIAMETER TABLE FOR PIPE AND TUBE (INCHES)

Nominal I.D.	CPVC Pipe*	Copper K	Copper L	Copper M	PB Tube	Steel WLS	Steel S40
0-3/4S	0.713	NA	NA	NA	NA	NA	NA
0-3/4	0.884	0.745	0.785	0.811	0.715	NA	NA
1-0/0	1.109	0.995	1.025	1.055	0.921	1.087	1.049
1-1/4	1.400	1.245	1.265	1.291	1.125	1.426	1.380
1-1/2	1.602	1.481	1.505	1.527	1.329	1.650	1.610
2-0/0	2.003	1.959	1.985	2.009	1.739	2.125	2.067
	C = 150					C = 120	

NA - not applicable
*I.D.s based on document G-82A published by B.F. Goodrich.

Hazen-Williams formula (psi/ft):

where f.l. = friction loss
Q = flow rate
C = roughness factor
d = inside pipe diameter

$$f.l. = \frac{4.52 * Q^{1.85}}{C^{1.85} * d^{4.87}}$$

FRICTION LOSS IN FITTINGS
(equivalent feet of pipe)

	Nominal Size	ell-90	T-branch	T-run	ell-45
Copper	0 3/4"	3.0	4.5	1.5	1.5
	1 0/0"	3.0	7.5	3.0	1.5
	1 1/4"	4.5	9.0	3.0	1.5
	1 1/2"	6.0	12.0	4.5	3.0
	2 0/0"	7.0	15.0	4.5	3.0
CPVC Pipe*	0 3/4" S	2.0	4.0	1.0	1.0
	0 3/4"	2.0	4.0	1.0	1.0
	1 0/0"	2.5	5.0	1.5	1.5
	1 1/4"	3.0	6.0	2.0	2.0
	1 1/2"	4.0	8.0	2.0	2.0
	2 0/0"	5.0	10.0	3.0	2.0
PB Tube	0 3/4"	3.0	4.0	1.0	1.0
	1 0/0"	3.0	5.0	1.0	1.0
	1 1/4"	4.0	7.0	1.0	2.0
	1 1/2"	5.0	8.0	2.0	2.0
	2 0/0"	6.0	10.0	2.0	2.0
Steel S40	1 0/0"	3.0	5.0	2.0	1.0
	1 1/4"	3.0	6.0	2.0	2.0
	1 1/2"	4.0	8.0	3.0	2.0
	2 0/0"	5.0	10.0	3.0	3.0
Steel THW	1 0/0"	3.0	5.0	2.0	1.0
	1 1/4"	3.0	6.0	2.0	2.0
	1 1/2"	4.0	8.0	3.0	2.0
	2 0/0"	5.0	10.0	3.0	3.0

*Based on data published by Central Sprinkler, 3/4-S estimate.

FRICTION LOSS FACTOR TABLE (psi/ft)

NOMINAL SIZE	FLOW RATE (GPM)											
	10	12	14	16	18	20	22	24	26	28	30	32
3/4" CU (M)	0.08	0.12	0.16	0.20	0.25	0.30	0.36	0.42	0.49	0.56	0.64	0.72
1" CU (M)	0.02	0.03	0.04	0.06	0.07	0.08	0.10	0.12	0.14	0.16	0.18	0.20
1 1/4" CU (M)	0.01	0.01	0.02	0.02	0.03	0.03	0.04	0.04	0.05	0.06	0.07	0.07
1 1/2" CU (M)	0.00	0.01	0.01	0.01	0.01	0.01	0.02	0.02	0.02	0.03	0.03	0.03
2" CU (M)	0.00	0.00	0.00	0.00	0.00	0.00	0.00	0.01	0.01	0.01	0.01	0.01
3/4" CPVC	0.05	0.08	0.10	0.13	0.16	0.20	0.24	0.28	0.32	0.37	0.42	0.47
1" CPVC	0.02	0.03	0.03	0.04	0.05	0.07	0.08	0.09	0.11	0.12	0.14	0.16
1 1/4" CPVC	0.01	0.01	0.01	0.01	0.02	0.02	0.03	0.03	0.03	0.04	0.04	0.05
1 1/2" CPVC	0.00	0.00	0.01	0.01	0.01	0.01	0.01	0.02	0.02	0.02	0.02	0.03
2" CPVC	0.00	0.00	0.00	0.00	0.00	0.00	0.00	0.01	0.01	0.01	0.01	0.01
3/4" PB	0.15	0.22	0.29	0.37	0.46	0.56	0.66	0.78	0.90	1.04	1.18	1.33
1" PB	0.05	0.06	0.08	0.11	0.13	0.16	0.19	0.23	0.26	0.30	0.34	0.39
1 1/4" PB	0.02	0.02	0.03	0.04	0.05	0.06	0.07	0.09	0.10	0.11	0.13	0.15
1 1/2" PB	0.01	0.00	0.01	0.01	0.01	0.01	0.01	0.02	0.02	0.02	0.02	0.03
2" PB	0.00	0.00	0.00	0.00	0.00	0.00	0.00	0.01	0.01	0.01	0.01	0.01